CAMBRIDGE ANTIQUARIAN RECORDS SOCIETY

VOLUME 3
FOR THE YEARS 1974 & 1975

THE WEST FIELDS OF CAMBRIDGE

Edited by
CATHERINE P. HALL, M.A.
and
J. R. RAVENSDALE, M.A., Ph.D.

IN PIAM MEMORIAM
MATTHEI PARKER

CAMBRIDGE
1976

© Cambridge Antiquarian Records Society
ISBN O 904323 02 1

This volume has been published with the help of generous grants from
The University of Cambridge
Corpus Christi College
Gonville and Caius College
Saint John's College
The Twenty-Seven Foundation Awards

Type set in Times New Roman, printed and bound at Pendragon Press, Papworth Everard, Cambridge

AUTHORS' PREFACE

"It well deserves to be carefully edited." In 1898 Maitland called for an edition of this Terrier, and it seems a fitting act of piety to his memory to answer that call before its centenary, and to tackle at least the main manuscript source for the work which he hoped would "restore the defaced pattern of the Cambridge open fields."

Maitland used this document when compiling the lectures later printed as *Township and Borough*. There he acknowledged his debt to Seebohm, and pointed to Seebohm's extensive use of it: "This terrier supplied Mr. Seebohm with some of his best materials when he was expounding the open field." So it has the extraordinary honour of serving as foundation both for *The English Village Community* and *Township and Borough*. It may reflect the rural quality of medieval towns in England that such a double use was possible, but it may also be that the exceptional nature of the lands set out, town fields rather than merely village fields, has not received due allowance, and thus the seeds of error may have been sown in the pioneer work on the economy of the medieval English village. It is quite certain that the neglect of this terrier since Maitland's time, and the lack of an edition, has allowed a great deal of fierce academic controversy to be waged wastefully and indecisively during the past twenty-five years.

It is therefore our hope that the appearance of this edition may in some matters help to set this record straight, but more particularly it should serve to reveal how many aspects of local, agrarian and social history remain to be properly explored with the aid of the multitude of contemporary documents still awaiting attention here in Cambridge. In our commentary we have touched briefly on some of these subjects, necessarily tentatively and from a limited range of additional sources. If our work does no more than to incite others to go out and do much more and much better, we shall be more than satisfied.

C.P.H.
J.R.R.

CONTENTS

	Page
Authors' Preface	v
List of Maps	ix
Acknowledgements	xi
Abbreviations and Short Titles	xii
THE CORPUS TERRIER AND THE FOUNDING FATHERS	1
APPEARANCE AND PROVENANCE	6
THE DESCRIPTION OF LAND	12
Selions, gores, butts and cuttings; the units of ploughing.	13
Headlands	16
Strips and doles: the units of tenure	18
Balks great and small	19
Furlongs	25
Fields and seasons	26
Enclosures	28
THE MAP OF THE WEST FIELDS	
The area of the West Fields: boundaries, routes and crosses	33
Topography and lay-out	37
Place and Field Names	43
The nomenclature of the great fields	48
HINTS OF ORIGINS: early forms; the great plough-up; Cambridge as a shire-town; the shift to the port	52
LORDSHIP AND OWNERSHIP	
Lordship in the West Fields: A. Merton	57
B. Mortimer and St. John's Hospital	64
Land ownership	72
Tithe ownership	76
THE GENESIS OF THE BACKS	80
THE TEXT	88
APPENDICES	
A Topographical Notes in the Terriers of St. John's College	139
B Mortimer documents from Gonville and Caius College	144
C Trees in the West Fields	147
D Place and Field Names	149
E Tables of Ownership	156
INDEX	161

LIST OF MAPS

Page

Readers should be warned that the maps are inevitably composite reconstructions

I MIDDLE FIELD. Details from the Corpus Map of 1789 showing College possessions (numbered) with balks. Note the long selion and balk of Roger de Harleston, and the balk running from the end of Endlesse Way which marked the division of the Seasons from the seventeenth century. 20

II COTON BOUNDARY. Shows the complicated pattern of ploughing in the far west of Middle Field, as described in the Corpus Terrier. Early enclosure, shown by hedges in the large map, took place in this area. Note the broad balk separating the parallel selions of Coton fields. 40

III THE APPROACH TO THE TOWN FROM THE S.W. Part of Little Field, showing ownership, largely to Hospital and Mortimer in blocks. Note the realignment of the Roman Road (Barton Way). 71

IV NEWNHAM. The part of Carme Field containing Newnham Crofts. Names, houses and parish boundaries from various sources before 1800. 77

V THE SITE OF THE BACKS. Reconstruction of the portion of the Carme Field adjacent to the Long Green from the data given in the Terrier. 83

VI THE WEST FIELDS. To the outline taken from the Corpus map of 1789, the following have been added: Furlong numbers from the Corpus Terrier, with the alternative numbering used in later terriers for Grithow Field in Os. Extensions of the 1789 map in far E. & W. from the Draft Enclosure map of 1803, modified to correspond with data from the Corpus Terrier. Hedges, clay and gravel pits from the 1789 map. Field names from the Corpus Terrier and its successors. Miscellaneous symbols as in the text and margins of the Corpus Terrier. Rough indication of 50 ft. and 75 ft. contours for guidance. . . . . . *fold out between inside back cover*

ACKNOWLEDGEMENTS

The authors and the Antiquarian Records Society are grateful to the following for access to documents, and for permission to transcribe and to quote: Cambridge University Library for the text of MS. Add. 2601, *Terrarium Cantabrigiae;* the Master and Fellows of Corpus Christi College, Cambridge, for the map of 1789, terriers and numerous charters; the Master and Fellows of Gonville and Caius College, Cambridge, for charters; the Master and Fellows of Jesus College, Cambridge, for terriers and for the Account Rolls of the Nunnery of St. Radegund; the Master and Fellows of St. John's College, Cambridge for terriers; the Warden and Fellows of Merton College, Oxford, for charters; the Syndics of Cambridge University Press for permission to use the quotation from *The Letters of Frederick William Maitland* and the numerous references to *Township and Borough*. In addition we would like to thank the librarians, archivists and staff of all these institutions for great personal kindness and Mrs Elaine Butt of Homerton College for preparing the maps for publication. Above all we would like to offer special thanks to Professor Bruce Dickins and Dr. R. I. Page for their long-standing interest in our work, and to Professor Dickins for reading the manuscript of the commentary.

ABBREVIATIONS AND SHORT TITLES

T&B	F. W. Maitland, *Township and Borough,* Cambridge, 1898.
EVC	F. Seebohm, *The English Village Community,* 3rd edition, London 1884.
OF	C. S. & C. S. Orwin. *The Open Fields,* 3rd edition, ed. Joan Thirsk, Oxford, 1967.
Arch. Hist.	R. Willis & J. W. Clark, *The Architectural History of the University of Cambridge,* Cambridge, 1886.
Bernwelle	J. W. Clark, ed., *Liber Memorandorum Ecclesie de Bernewelle,* Cambridge, 1907.
Ag.H.R.	*Agricultural History Review.*
Econ.H.R.	*Economic History Review.*
C.C.C.	Corpus Christi College, Cambridge.
G. & C.C.	Gonville and Caius College, Cambridge.
M.	Merton College, Oxford.

THE CORPUS TERRIER AND THE FOUNDING FATHERS

"I am enjoying myself thoroughly over a terrier of the Cambridge common field. I think of making it the basis of my lectures in Oxford." Thus Frederick William Maitland to Henry Sidgwick, in March 1897.[1]

The lectures in which he carried out his expressed intention were the Ford Lectures given in that same year under the title, *Township and Borough*. They were published in the following year 'with an Appendix of notes relating to the History of Cambridge', an appendix which exceeded the lectures in length and depth of detail.

In his fourth lecture Maitland describes the document which we have taken for our text as follows:

"Now of these western or transpontine fields we have in our University Library a field-book or terrier. It was made to all appearances soon after the middle of the fourteenth century. It is the most elaborate thing of its kind that I have ever seen. In each field it describes the various furlongs or shots in such a manner that an ingenious man, who had time to spare and taste for the Chinese puzzle, might depict them on a map." [2]

The document consists of thirty-six folios of parchment, well bound in soft leather and boards, with an outer wrapper held by a brass clasp. As a result it is in an excellent state of preservation.

The original is written in a firm, clear hand, but the Latin is highly contracted: this seems to have caused the difficulty that the Orwins found when they consulted it. Maitland's suggested date of about 1360 is consistent both with the main hand and with the content. The acreage of the strips, two short terriers of Corpus lands in Chesterton, and many additional notes are in various later hands, most of which we think we can recognise from their use in other documents in the archives of Corpus Christi College, Cambridge.

The book, which would appear to have come into the possession of Corpus Christi College about the end of the fourteenth century (perhaps to replace documents destroyed in the Peasants' Revolt of 1381), was certainly still with the College two centuries later, and we have no reason to believe that it strayed much before the nineteenth century. Henry Bradshaw, University Librarian, bought it in December 1878 from Joseph Mayer, the antiquarian and collector, who had himself got it from another collector, Joseph Clarke of Saffron Walden. He drew the attention of both Seebohm and Maitland to it and himself made a transcript of the first twelve folios (U.L.C. Add. 4228). After his death it came to the University Library at his request. To folio 4v is now attached this note in Bradshaw's hand, "This book (a terrier of the West Fields of Cambridge MS) is to go to the University Library as my gift. Henry Bradshaw. Augt 31, 1885".

1 C. H. S. Fifoot, *The Letters of Frederick William Maitland*, Cambridge, 1965.
2 F. W. Maitland, *Township and Borough*, Cambridge 1898. Hereafter *T & B*. p. 56.

It is now in the University Library as the *TERRARIUM CANTABRIGIAE* under the reference number, Add. 2601. In its old home, it is known to the Parker Library as the "Corpus Terrier". This practice we follow here, since the name, "Cambridge Terrier", used in the University Library, applies equally to the various later versions in the archives of other colleges.

In form it is very like the village Field Books and Surveys which are so familiar in early modern times. It lists all the arable in the Cambridge West Fields, field by field, furlong by furlong, giving the strips in sequence, and noting the number of selions, butts and gores in each. The original did not specify the acreage of the strips, except in a few cases, but this has been added later, first for Corpus lands and then for all strips, usually in the right-hand margins. Throughout it gave the tithe owner for each strip, and a note on the distribution of the tithe sheaves was included in the initial description of each field. The purpose of those who made the original survey may well have been mainly to record the tithing. It appears that there was once an earlier version among the Tithe Books of Barnwell Priory, a book often referred as to *Blackmore* or *Blackymore,* an old name for the West Fields which survived until the late eighteenth century as the title of an estate map, made for Corpus Christi College in 1789.

The form of some of the later terriers in other colleges suggests that they may be hybrids from the two early versions. Where final proof of title is necessary, the Corpus Terrier, no less than the later ones, refers back to the Barnwell Tithe Books. The obsession with tithe continues throughout the series.

The mid-fourteenth century might be thought of as an early date, compared with the dates of most surviving village field books but it is late enough to make Cambridge West Fields somewhat dangerous ground for generalisation about the English medieval *village.* The document is atypical of village affairs not merely in that it describes town fields, but also in that it is concerned with the fields of the town of Cambridge, which had complex dual, if not triple, origins and another set of fields across the river. Whereas a village most often lay within the bounds of a single parish, it was meaningless, as Maitland pointed out, to talk of bounds of the Cambridge parishes in its open fields where the tithe allocation suggested a distribution on quite other principles, and one which moreover bore signs of having undergone considerable revision.

Both Seebohm and Maitland appear to have studied the document carefully, although Seebohm, wearing his learning lightly, describes it as, "a manuscript terrier of one of the open fields near Cambridge, belonging to the later years of the fourteenth, or the beginning of the fifteenth century." He describes its contents simply but accurately:

"It gives the names of the owners and occupiers of all the *seliones* or strips. They are divided by balks of turf. They lie in furlongs or *quarentenae.* They have frequently headlands or *forerae.* Some of the strips are gored and called *gored acres.* Many of them are described as *butts.*"[1]

He goes on to comment on symptoms of change:

"But this terrier also contains evidence that the system was even then in a state of decay and disintegration. The balks were disappearing, and the

1 F. Seebohm, *The English Village Community* (3rd edn.), 1884. Hereafter *EVC.*

strips, though still remembered as strips, were becoming merged in larger portions, so that they lie thrown together *sine balca*. The mention is frequent of iii.seliones which used to be v., ii. which used to be iv., iii. which used to be viii., and so on. Evidently the meaning and use of the half acre strips had already gone."[1]

Seebohm failed to distinguish consistently between the unit of ploughing, the selion, and the unit of tenure, the strip, which more often than not contained more than one selion. But consistency in the use of the modern term 'strip' has only been demanded and defined of very recent years, and we have no right to blame Seebohm for not arriving at it a century ago. Otherwise, so far as the contents of the terrier go, Seebohm's was an admirable factual description. In one small particular only, the statement that the balks consisted of strips of turf, was he going beyond the evidence contained in the document itself. The correctness of the assertion has to be proved from supplementary material. The evidence of change that he cited is very clear, but in its interpretation we cannot expect Seebohm to have shared the benefit we ourselves have of Professor Postan's delineation of the course of economic change in the middle ages.

In *Township and Borough*, after praising Seebohm's work, Maitland carried the study and exposition of the terrier further, giving us a sample translation of a few entries for Grithow Field[2] This we would only wish to revise in one respect his rendering of the word *iuxta* as "about". There are numerous entries for which he could not possibly have translated it so, and he makes it qualify the figures of acreage which were not entered in the Ms. until decades after the *iuxta*. The translation was probably made from notes in his study away from the original, though he recognised elsewhere that the acreages were in another hand. The later terriers, when they are in English, clearly show the correct meaning of the term as "next", merely indicating that the strips are following in sequence. Part of his sample is as follows:

"1 selion of Thomas Bolle late of John Toft about [1 rood]—Radegund."

Had he continued a little further on the same page he would have had to use a different form of translation:

"j selio Thome Bolle quondam Johanis de Toft iuxta et est longior aliis selionibus antedictis et in parte forera ad capud suum boriale [di acre]— Radegund." (fol. 1v.)

There are a number of entries, where one sequence of strips has ended and another begins where *iuxta* does not appear, but acreage is still given. For a fair proportion of the entries Maitland's translation of *iuxta* as "about" is quite impossible. "Next", as in the later English terriers, always makes good sense. Any omission of *iuxta* means that the strip does not lie alongside its predecessor and a new sequence begins.

Maitland fastened on some of the differences between this and some of the other field books which he knew, and with characteristic brilliance related these differences to the origins and development of the borough of Cambridge. There were no signs of regular hides and virgates: documents cannot take us back to villeinage in the Cambridge fields. The system of tenure in the fields denied both

1 *EVC*, pp 19–20.
2 *T & B*. p. 124.

feudal and manorial origins: the "manors" found in later Cambridge were all later creations. The few copyholders, all of Merton, discovered in the later days of the West Fields were no descendents of villeins. He saw signs of great activity in the transfer of land in the years before the Corpus Terrier was written, involving country landowners as well as rising and falling burgensic families. By 1360 the wave of endowments of religious houses was being followed by a wave of endowments to Colleges. In his notes he set out the topography of the fields as he found them in the terrier, his provisional tabulation of lands in both East and West Fields, and his conclusions showing how the changes in the fields related to social change in the borough. Characteristically he opened up a great many of the questions to which we are still seeking an answer today. There can be little doubt that more progress would have been made if the text of the document he used had been easily available.

When writing *The Open Fields*[1], the Orwins took a glance at the Corpus Terrier to refute Seebohm's contentions about balks. Unfortunately their glance was not penetrating or long enough, and subsequent scholars have been deterred from tackling abundant evidence in this source because they interpreted the Orwins' remarks to mean that Seebohm had invented his evidence. The Orwins said:

"A recent examination of the manuscript revealed, in fact, that balks are seldom mentioned, and then only as important boundaries. There is no indication that they divide strips, and the phrase *sine balca* could not be found."[2]

Alas! the sad truth, and it is a sad truth to find error in such a splendid work, is that the phrase *sine balca* occurs in the left-hand margin (albeit in contracted form since all the text is in highly contracted form) no fewer than fifty two times. It appears first on the verso of the first folio, describing the second furlong of Grithow field, coming shortly after the alternative phrase, *inter duas balcas*, which occurs more than twice as often as *sine balca* in the text. It will be shown later, in the analysis of the terrier, that an enormous number of balks which separate strips is implied by the text, that the implication was meant, and that most of them survived up to the eve of Parliamentary Enclosure in the nineteenth century. The Orwins' failure to spot this has effectively buried what was perhaps the best documentary evidence to come to light yet for examining the vexed problem of balks in the open fields. Dr. Kerridge's suspicions were not aroused by the hesitant tone of the Orwins' reference to "inspection": "The document Seebohm referred to was found to mention only a few balks."[3] In the summary of the controversy and evidence in the *Agricultural History Review* of 1956, " A Review of Balks as Strip Boundaries in the Open Fields", no evidence is recorded for Cambridgeshire.[4]

1 C. S. and C. S. Orwin, *The Open Fields* (3rd edn.,' ed. Joan Thirsk), Oxford, 1967. Hereafter *OF*.
2 *OF*, p. 44.
3 E. Kerridge, "A Reconsideration of Some Former Husbandry Practices", *Ag. H. R.* iii, no. 1, 1955.
4 H. A. Beecham, "A Review of Balks as Strip Boundaries in the Open Fields", *Ag. H. R.* iv, no. 1, 1956.

The Corpus Terrier gives direct evidence of only one set of open fields, and a peculiar town set at that, but there, in combination with other documents, it sheds far more light than is often possible for any field system. We can follow the detail and pattern of the Cambridge West Fields from a base in the fourteenth century right through to the nineteenth century. We have also been able to convert the words of the fourteenth century scribe into an exact map as Maitland hoped. This would not have been possible from the terrier alone, simply by solving the Chinese puzzle, as Maitland called it, but the First Draft of the Enclosure Award Map in the County Record Office, and the discovery of a late eighteenth century map of the College lands in Blackmoor Fields in the Bursary of Corpus Christi College brought the task well within the limits of human patience.

The result is even more complicated then expected, and some of the peculiarities may be unique to this set of fields. But the combination of Terrier and map, as well as restoring the vanished face of the old landscape, gives us substantial evidence on such matters as the relation between medieval selions and modern ridge and furrow; selion size and changes in selion size; balks of different kinds and the ploughing up of balks; the relation of traditional measured acres to exact acreages surveyed in the eighteenth century; the relation of a three shift system to the topography of a set of fields numbered as four and on occasion six. And again we find that in another set of Cambridgeshire fields, a period of change before the late fourteenth century was followed by something like ossification of their pattern.

We should feel well rewarded if our work added its little weight to the rehabilitation of Seebohm's reputation. Maitland freely acknowledged his debt, and claimed to be Seebohm's disciple in the study of field systems. There has been a recent revival of interest in Seebohm's speculations as he attempted to break through the barrier of the unknown in the Dark Ages to the problem of survival from Roman Britain. It is perhaps ironical that his firmer steps in what to him was the known, English open fields as revealed by this terrier, should have been neglected as errant. Seebohm may have been unfortunate in that his generalisations drew heavily on what proves to be an eccentric example, Cambridge. But it may still be a wiser policy to learn what we can from this peculiar place, rather than to ignore its existence. Through four and a half centuries the life of the Cambridge West Fields can be revealed in great detail and clarity. Such a study should have value in providing examples on matters in which they were not peculiar, and also, by the identification and explanation of their peculiarities, teach us something more of the medieval town.

APPEARANCE AND PROVENANCE

In appearance the Terrier is in book form of a handy page size $8\frac{1}{2} \times 5\frac{1}{2}''$ (21·5 × 14cm). Its thirtysix folios are written on tough vellum in three gatherings, stitched onto four sheepskin bands and bound between oak boards. A contemporary sheepskin wrapper gives additional protection to a document that was expected to be carried round and handled a good deal.

The first and last leaves of the book are too clean for it ever to have been used unbound. It is probable that in its original state it was bound straight into the bevelled, uncovered oak boards, as its counterpart terrier of the East Fields[1] is to this day. At some point in the later middle ages, probably the mid to late 15th century, two sets of flyleaves were added fore and aft of the original text. These consist each of four leaves approximately twice the size of the terrier leaves, trimmed with a knife and folded to make four end-papers. They come from a fifteenth century *"Legenda Sanctorum"* written in double columns on white vellum of rather better quality than that of the Terrier, but every page has some slight fault or is incomplete, suggesting that it is the spoil of a scriptorium being used up by a contemporary binder. The earliest notes written on the flyleaves are in hands dating from late fifteenth to mid-sixteenth centuries. We think that it was at this time that the oak boards were covered with sheepskin, which is again of rather superior quality to that of the carrying wrapper.

At some later stage also the first and last of the flyleaves were pasted down onto the oak boards in the manner of end papers. This practice was not normal before the sixteenth century. Its results were sometimes unfortunate, if only because the paste seems to have attracted beetles. Here in Ms 2601 beetles have started burrowing where the end fly meets the cover and have gnawed a certain amount of flyleaf here and there. Fortunately they never penetrated to the main text.

From the sixteenth century to the present day the manuscript has retained substantially its present appearance, with *Occidentes campi* in black ink on the outer wrapper. Close examination reveals that it has had one very skilled re-sewing. This has been done by the method of cutting back the cover over the spine sufficiently to insert a new sheepskin band like a headband, head and tail. This method, which was extensively used in the eighteenth century, has the advantage of leaving existing thongs intact. This re-sewing could have been done in the eighteenth century, perhaps under Robert Masters, or in the nineteenth by a binder conversant with older methods. To the period of the re-sewing we would assign the leather strap and pin which hold the sheepskin cover in place.

The text of the Terrier is neatly set out in a single main column of 27–30 lines per page with sufficiently wide margins to allow for marginal entries on either side. These might almost be said to form narrow columns in their own right. Headings are clear and usually in rather larger letters. The main text is in a style

1 C.C.C. *VII.*

of handwriting that was current in the late fourteenth century for better-class charters and administrative documents, found in University and College muniments. At times the headings include flourishes more common in the early fifteenth century, but Cambridge seems to have been quick at adopting the latest fashions in handwriting. The layout and style all suggest a professional scribe, academic or monastic trained, of the latter half of the fourteenth century.

Another point about the layout is that the leaves are arranged in three gatherings of twelve leaves, and the three main sections of the text, relating to the three big fields, begin on fols. 1, 13 and 27 respectively, that is to say, on the first folio of a gathering. But the middle section slightly overshoots its quota, so that the third section starts on fol. 27b. This suggests that the scribe had a pretty good idea of the amount of space he needed for his text in advance, most probably that he was making a neat copy from an original of similar format.

It has already been indicated that the document called by Bradshaw, *Terrarium Cantabrigiae*, is so obviously from its layout and headings primarily concerned with tithe that it might more properly be called the Cambridge Tithe Book of the West Fields. Moreover references in the text to "diversos libros de decimis de Bernewell" make it clear that it was written by or for people who had normal access to the Barnwell Tithe Books in current or very recent use. We would therefore infer that the text we have is derived from an original in Barnwell and may well have been made by, for or in, Barnwell itself. Whether our "Corpus" Terrier is a unique copy, or one of several contemporary copies of the same text we have no means of knowing, only that it is one of a family of texts derived from a common source.

Internal evidence enables us to date our text with some precision. The appearance of Corpus Christi College puts the date as after 1352. The omission of the parish of All Saints by the Castle from the tithing system and the very large share received by St. Giles (EGID) puts the date after the union of parishes on Castle Hill, which followed the depopulation caused by the Black Death, namely in 1365. It is probable that the reorganisation of tithe collection necessitated by the absorption of All Saints also caused the writing of a revised version of the existing Barnwell Tithe Book. The version we have in the Corpus Terrier gives names of contemporary owners which indicate that it must have been written in the last years of the reign of Edward III at the latest.

For instance, a quitclaim of 1365[1] was witnessed by the following, who appear as owners or 'quondam' owners, in the Corpus Terrier:—

    Stephen Morice, junior   Richard de Arderne
    Thomas Bolle   John Gyboun
    Richard Martyn

An earlier deed made in 1349[1] has the following 'quondam' owners as parties or witnesses:—

    William de Lolleworth   Richard de Arderne
    Henry de Beche   John Gyboun

If we accept the suggestion that where "quondam" owners are named, the change of ownership is recent (i.e. since the writing of the Barnwell original) while unchanged owners and owners of long standing are entered without a

1 C.C.C. Miscellanea XXIV B. 38, 34.

"quondam", we can postulate that the writer of our text was working from a version of c. 1340 or at least one kept revised to that date.

After the writing of the "revised version" given in the Corpus Terrier, however, this version took on the nature of a *textus receptus* and all known copies or partial copies found in St. John, Jesus, Clare and Corpus follow its form and its quondam owners. So the writer of the amendment slip found with fol. 5 of the Corpus Terrier describes a selion as "quondam Thome Bolle (14th c. owner) nunc Mr. Wood (16th c. owner)," ignoring all in between. Maitland refers to an even more extreme case of the same thing, where the nineteenth century owner, Mr. Panton, appears as if he were the immediate successor to the Prior of Barnwell.[1]

While the interest of the marginalia, notes and other later insertions in the text lies for most readers in the changes they record in the fields themselves (e.g. *nunc gravelpyttes, nunc in clausuris* etc.) they can offer fascinating clues as to who owned and used the Terrier at various dates.

The earliest of the marginalia would seem to be those letters and symbols in the margin which would enable an official to spot at a glance the selions belonging to certain owners. These owners appear to be the Nuns of Beche (Waterbeach) (marked with a "*b*"), the White Canons of Sempringham ("al ca"), the Clerks of Merton ("mer"), and are in hands stylistically similar to that of the text of the Terrier itself. Almost contemporary or only very slightly later, in a pale ink, is the "mo" against the lands of the Morys family, usually but not exclusively those of Stephen Morys senior. From this we might guess that the first users of the terrier were officials concerned with the collection of dues for the Nuns of Beche, the clerks of Merton and the White Canons. These were all corporate bodies with lands and tenements in Cambridge and may well have employed the same collector. The presence of the "mo" is more obscure, since the Morys family owned so much property locally that they would surely have had their own bailiff and receiver. It could be that somebody was interested in assessing the extent and value of the Morys holdings with an eye to purchase or bequest. Alternatively they could be merely instructions to a copyist making a terrier or partial terrier for the use of the Morys family. Such partial copies for other owners (e.g. the University[2], Clare College[3]) still exist.

None of these early marginalia suggest any connection with Corpus Christi College. Indeed, the *cχ*, the most striking and frequent of all the marginal symbols, is in a fifteenth century type of hand. So is the *nota,* put to catch the eye every time the tithes of St. Botolph are mentioned, and the summary of tithing and selions of St. Botolph's, added at the end of the section relating to each field. These two sets of notes are similar in style and hand to various notes and endorsements on the Corpus Christi College muniments made during the Mastership of Dr. John Botwright (Master 1443–1474). It was in 1460, during his mastership, that the status of St. Botolph's was restored to that of a rectory and several other legal points cleared up, prior to its exchange with the newly founded Queens' College for St. Bernard's Hostel.

1 *T. & B.,* p. 55.
2 In C.U. Archives, 'University Lands'.
3 In Jesus College.

APPEARANCE AND PROVENANCE

At about the same period, to judge from the dark ink used for both, the marginal sketches of the crosses, the conduit and the grithow were gone over, and the $c\chi$ s were elaborated with additional dots and letters, including a most confusing second "b" (distinguishable only in style from the "b" of the Nuns of Beche). Similar combinations of letters and dots are found on the dorse of some of the Corpus deeds, and it would appear that they are a sort of filing code to enable the bursar or clerk to lay his hands on the right bag or small box within the main College muniment chest. The University had a similar though more elaborate system by 1500. All in all it is clear that the Terrier was very much in use in Corpus in the fifteenth century.

When and why did the Terrier come to Corpus, if it had not originally been produced for the use of that College? One suggestion, made before we had gone into the internal evidence of the marginalia, was that it might have been acquired as a replacement for Corpus documents lost in the uprising of 1381. The oldest extant copy of statutes of the University still in its possession has long been believed to be such a replacement. This suggestion appears even more reasonable when we come to trace the hands of the unknown writers who added acreages and other topographical notes to the Terrier.

The first of these writers confined his additions of acreages to the Corpus properties and did so in a bold scribbly hand, which created difficulties for his neat successor who was trying to fill in all the acreages and was not always left enough space. This alone is enough to show that the Corpus acreages were added first, probably by the man who first acquired the book for the College's use. This same bold hand is found in the College muniments in the rentals of John Brunne, receiver of rents for the College in the reign of Richard II, and we have every reason for believing that the hand is his. The second neat hand is also found in the College muniments on various documents, including one connected with John Brunne. This is a lease to Brunne of two houses in Bridge Street from the College (Richard Billingford, Master) and dated 1399.[1] On the dorse of this lease is a summary list of owners and acreages in Portfield and Carmefield in the hand of the tidy writer. The names of the owners would place it as about one generation away from the owners in the Corpus Terrier. (Some have the same names, some only the same family name, some differ altogether). It is, therefore, the hand of a man who was active in the affairs of Corpus at the turn of the fourteenth–fifteenth century, possibly that of the Master, Richard Billingford, whose brass is in St. Benet's Church.

Possibly in the same hand, or one of very similar neatness, is the marginal "clar' hall" against the possessions of John de Weston. If it is the same, it would help to fix the date at which Clare College acquired the former lands of John de Weston, a date at present not known, owing to the destruction of the older Clare records by fire.

At the back of the Terrier, on the blank pages left at the end of the Carmefield section, is a partial terrier of Corpus properties in the Chesterton Fields, added in a rather dashing 15th c. hand (fol. 35). Of about the same date are some of the earliest notes found in the flyleaves, (notably fols. 1. & 37) fol. 36 is taken up with a copy of the same Chesterton terrier as that on fol. 35 and this one is

[1] C.C.C. *XVI*.B. No. 3a.

fortunately dated for us; it runs,
> "renovatur per Petrum Nobys clericum custodem dicti collegii anno domini 1517° et anno regni regis Henrici octavi IX° sexto die Octobris."

In a similar and contemporary hand are the topographical data which form a notable feature of fol. 1 and of all the Carmefield section. The writer of these, notes appears to use English and Latin impartially. At about the same time some revision of the names of tenants was made (e.g. "griffyn" on fol. 20v) From Gonville and Caius College muniments we can date the addition of "Bokenham" whose lands came to that college in 1524.[1] The "gn" in a neat hand marks the acquisition of the Mortimer manor in the same decade.

When we come to the Mastership of Matthew Parker, in Corpus Christi College, we are not surprised to find evidence of careful checking under that meticulous administrator. The College archives contain a pair of draft fieldbooks, summarising the College lands in the West Fields which go with a particular farming lease, and dated 1549. The Terrier bears a note on the third flyleaf, in the same hand as these draft books, which relates various marginal signs to previous tenancies;

> memorandum quod in campis sequentibus super diversas seliones pertinentes collegio nostro; ubi scribentur littere l. vel lu: f. et Elis signat tenentes earum terrarum que sunt in tenura J. Stephani nam J. Fynne R. lucas et Th. Elis fuerunt firmarii eiusdem ut in registro fol. 8°
> Item sillaba "ca" signat item alias terras quondam Wm. Cap.
> Item sillaba "co" signat illas terras que sunt in tenura Johannis Cole de Barton.

There is another note in the flyleaves (iv) in a hand not as yet identified.
> Slegge habet in medylfelde 3 quarent' di acr' 35, di acr' iniuste.

Systematic amendments and annotations to the Terrier cease about the midsixteenth century. That the Terrier was still in Corpus at the end of the 16th century, even if not normally in current use, is shown by a slip of paper left loose in the Terrier which relates to an amendment on fol. 5v, indicated by a pointing finger. The amendment describes in detail the four selions "quondam Thome Bolle" which form a small furlong by themselves and are now in the possession of Mr. Wood, the wealthy landowner of Fulbourn. Scribbled below it is the following, "this note appearethe in all other terriers savinge this of the Collidge compared a° 1594. Eliz 36°." From this point onwards the history of the Terrier can only be conjectural.

Comparison of the Corpus Terrier of the West Fields with the similar volumes for the East Fields still in possession of the College[2] is a help in reconstructing their use and history. Both these books, one a 14th century one and 15c. copy of the same text, bear many of the same marginalia as the West Fields Terrier. Here is the very familiar *c%*, the attendant dots and signs for internal filing, and *nota*. As might be expected, the 15c. copy is more fully annotated and here we also find "mo" against the lands of Stephen Morys senior. The volumes are not in origin companion volumes to the West Fields Terrier, but have obviously been used by the same people in Corpus, and for similar purposes. If anything,

1 G. & C.C. *Registrum Magnum*. p. 279.
2 C.C.C. *XIV & XVII* 5.

the East Field Terrier appears to have gone out of regular use at an earlier date, to judge from marginalia alone. But it bears a note on the first flyleaf which is very relevant to the possible subsequent history of the West Field Terrier. This, in a bold mid-16th century hand, runs "m^d that matthwe parker a° Dni 1552 20 Julii et a°r.r. Edwardi sexti sexto deliveryd a newe wryten terry of the East feld copied out bi this to Wylton Robynson fermer to the college of the birdbolt in Barnewell and fermer of the porcion longing to benet paryshe to his use, but to be restored to the college of corpus christi aforesaid. In presentes of Andrew person bachiler of dyvynytie and Roger yong bayly of the said college".

It would seem reasonable to suppose that if Matthew Parker had a neat copy of the East Fields Tithe Book or Terrier made for College use, he also had one made for the West Fields. As the College's farm holdings were greater in the West Fields, though their tithe was less, it is possible that only partial terriers, similar to the rough drafts of 1549 already referred to, were made for the West Fields. But in any case, the existence of neat working copies of what was needed for current use would explain why the older terriers became obsolete and ceased to be annotated.

In the better copy of the East Fields Terrier (XVII 5) there is a second note on the first flyleaf, below the one already quoted. This is by Robert Masters, the eighteenth-century historian of the College who made the first catalogue of its muniments. He has written, "The Book described above, with *Mr. Robynson's* name in it is now in the possession of *Jacob Butler* Esq. owner of the Priory of Barnwell. R.M." Moreover this old East Field Terrier has also been restored and re-sewn in the style described above which expert bookbinders say is characteristic of eighteenth-century repair. The boards of this volume, unlike the other, cruder copy, are covered with undressed white skin and there has relatively recently been a clasp. Now the West Fields Terrier has been re-sewn and repaired in exactly the same style, and a solid, hand-made metal clasp on a leather strap added. The similarity of repair on both Terriers of East & West Fields lends weight to the suggestion that both were in the College at the time of Robert Masters and both repaired at his orders. Vestiges of a drawer or catalogue number in Masters' hand appear on the dorse of the East Fields Terrier. In the West Fields's Terrier the flyleaves have been stuck down like endpapers. In the East Fields Terrier a rather unsuccessful attempt appears to have been made to do the same, only fortunately the paste has not held.

What happened to the West Fields Terrier between its repair in the eighteenth century and its appearance on the second-hand market in the nineteenth century has been and remains a mystery.

THE DESCRIPTION OF LAND

When Maitland wanted to show the difference between describing a piece of land in terms of the ploughing-unit, the selion, and describing it in acres and roods, the units of computation, he wrote, "the selion is the visible fact, stamped upon the face of the land; the acre, on the other hand, is a matter of traditional reputation". The same sort of mental distinction is necessary when we come to consider furlongs, for we run up against the difficulty that the same word is used to cover two related but different things. In its simpler meaning it is a practical unit of length, the distance the oxen could plough one furrow before turning, the length of a selion. But it is also used of a two-dimensional unit, a row of furrow-long selions lying side by side. On a detailed map or in an aerial photograph it is noticeable that the pieces are by no means all and always the same length. Nor is there any standard or specific width to a group; they may widen into a bulge, tail off into a point, or stop and be met by another bundle of selions lying transversely. These are the furlongs of fact "stamped upon the face of the land", and their only common characteristic would seem to be that (with exceptions), they are just one furrow long.

The Corpus Terrier, in common with most administrative documents of the same period, is in Latin. When describing a ridge of land the word *selio* is used, and in spite of the plethora of alternatives in the local vernacular this word got adopted into English as "selion". The word used in the Terrier, and more generally in this part of the country, to translate "furlong" is *quarentena*. Forty rods theoretically make one furlong, and superficially the term is a simple translation, 'a group of forty'. But when we move into Latin our thinking seems to change, and we are in another world. *Quarentena* is a technical term, devised by an alien administration using a foreign language, and one of a number of such administrative terms used in documents from the Conquest and through out the Middle Ages, which simply do not fit the facts that the administrator is trying to set down. *Quarentena* is an administrative concept, rooted ultimately in a civilisation which, ideally, made all its roads straight, cut its watercourses similarly and ploughed with straight furrows; from which it followed that its arable fields were, ideally, square and eminently measurable.

Worked out on the ground it would have produced a landscape reminiscent of 'Through the Looking-Glass' where Alice jumped over a little brook and landed in the next square. But it is obvious that this Looking-Glass world of the administrator bore very little relation to what was actually found in the Cambridge West Fields, or most other fields in this country, for that matter.

In the actual Fields, quite other considerations governed the lay-out of the groups of selions known as furlongs. The contours, though gentle, could be not ignored; watercourses meandered diagonally across the area under cultivation; roads, even when Roman and straight, radiated like spokes of a wheel, and where not straight followed a rolling Chestertonian course. Moreover, the ox-ploughs moved in a curved path and turned short of a ten-chain furrow if hindered by contours, heavy soil, or existing obstacles. The ploughing direction was set to throw off surface water as rapidly as possible by responding to the natural contours and tilt of the land. Added to this was the fact that the men who laid out the Cambridge West Fields did not do so all at one time. Some parts were probably already cultivated when they settled, and the others were opened up piecemeal over several centuries, according to the expanding needs of the community. The result was the varied and diverse scheme revealed by the map—Maitland's "Chinese Puzzle".

The way in which the administrator had to wrestle with chaotic data to bring it in line with the traditional language and lay-out of his documents is shown in the Corpus terrier in a number of ways. Most curious of these is in the matter of the compass points. Latin only provides for naming the four cardinal points without great circumlocution, so that it was obviously convenient that the Huntingdon Road should be reckoned as running due North, though it in fact goes North-West. Since theoretical *quarentene* are rectangular, it follows that the next band of furlongs lying alongside those abutting onto the Huntingdon Road should be presumed to have the same orientation; and similarly the next band. So when one works round to the Madingley Road, in spite of the fact that the roads radiate from Castle Hill, enclosing an area that is a segment and not a rectangle, by the application of the logic described above the Madingley Road is deemed to run due North, though it actually runs almost due West. So the process continues down to the far corner of the Fields. Here, where what is being called North is in reality nearly South-West, common sense at last breaks in, and the directions are turned by a right-angle for the last segment. Some attempt has also been made to correct the orientation in later marginal notes for Dedale and Little Field.

This general point of the dichotomy between reality and its presentation in documents has been made at some length, because it would seem that some commentators have not fully appreciated the tremendous variety in shape, size and content of actual furlongs on the ground, and the near-impossibility of lining them up with any theoretical furlongs of computation, and the difficulty of translating written abuttals into a map. The map of the West Fields should act as a salutary check upon any attempts at reconstruction of a map based on mathematical data alone.

Selions, Gores, Butts, and Cuttings:
The Units of Ploughing

There was much that was typical of midland open fields in the general lay-out of these fields on the west side of Cambridge. They were ploughed in selions, with gores and butts in awkward places where the lie of the land distorted the normal selion shape. The distinction between these three forms of ploughland was not an exact one. Sometimes a selion was broader at one end than the other, and so was said to be gored, or described as a gored-acre. Sometimes it was shorter, and was given the local term, a cutting or a butt (fol. 5v.). If we include gores and butts with selions, and exclude headlands (except where they are also called selions), since some of them at least were not ploughed, we can count a total of 3376 selions in the Corpus Terrier. By the seventeenth century, the fields sometimes stopped short of the boundary, as later versions of the Terrier make clear, e.g. the entry at the end of Furlong 6 in Grithowfield reads:

> Thus far lieth Cam. field, but the bounds of Camb. lieth further into the Closes as doth appear in Blackamore.

The closes mentioned were those associated with Howes, an ancient[1] settlement which extended across the boundary into Girton.

Thus the total count of selions depends on decisions that are to some extent arbitrary, and could be varied up or down, but 3300 to 3400 is the right order for the number of lands in the West Fields in the fourteenth century.

As Seebohm pointed out, some of the strips are recorded as having had the number of their selions reduced: 62 have disappeared in this way. Seebohm suggested that this was recent when the terrier was complied, and he may well have been right. Although the process continued, the total reduction during the four centuries after the Terrier was written was much less. A high proportion of the recorded reductions of the 1360's are by one selion only in a strip, although in a few cases the reorganisation seems to have been more extensive. In Grithowfield one strip of eight selions had become one of four, and in Carmefield one of eleven selions had become seven.

This is not a feature peculiar to these town fields: in the field books and terriers of many of the nearby villages there are signs of selions having been thrown together. The common name in the villages where a pair of selions lying side by side had been ploughed together was a "broad", or a "broadel selion". Stretham Glebe Terrier has a descriptive variant of this term, "a twilstitch", and where selions were ploughed together end to end, calls them a "throughout".[2]

This second method of re-arrangement also appears to have been going on, where numbers of double length selions, which make furlong boundaries very complex, are said to "extend through two furlongs". This is obviously where both time and land is saved by not having to turn the plough. It may be found most often in the flatter parts of Middlefield. Sometimes the later versions of the terrier mention, in places where blocks of selions in one furlong abutted similar blocks under the same ownership in an adjacent selion, that they were ploughed as a single group. The Merton pieces in Dedale are a good example.

Seebohm interpreted these piecemeal amalgamations as signs that the system was already in decay and that the half acre strip had lost its meaning. But although the average selion size is near half an acre, many remain at a single rod. It is difficult to detect any uniformity behind the lay-out of the late fourteenth century. A curious entry (fol. 5v) underlines this difficulty: a selion of Robert Long is described as containing one rod, but the size is given in the margin as half an acre. Maitland observed that the selion size varied from a single rod to over one acre, but we have found selions as large as an acre extremely rare,

1 St. John's College, xxxi. 29, 1617.
2 J. R. Ravensdale, *Liable to Floods,* Cambridge (1974).

and have not noticed any larger than this. But nevertheless a selion can be used as a rough indication of size: the width of a croft is in one case (Thorpe's croft) given as four selions' breadth.

Gores and butts varied in the same way. In nearby Landbeach, where the selions were more commonly close to half an acre, the butt was reckoned as one third of an acre. But in the Cambridge West fields there seems no such standard, some butts being larger than the normal selion. The term here refers much more to shape than to size, as does gore.

Although the process of reduction in the number of selions by combination continued, some of the later terriers show examples of the reverse process of subdivision, although these are very rare. The net change in total after c. 1360 seems to have been slight.

The Map of Blackmore Fields of 1789 shows all the strips of Corpus Christi College, and its schedule gives the measured acreage of each such strip and the number of ridges which compose it. When we make allowance for the occasional throwing together of selions, this confirms the ridges of the modern period as the descendants of the selions of the fourteenth century. This point was argued by Beresford from the comparison of air photographs and strip maps.[1] In Cambridge we also have the evidence of surviving earthworks to help us. The University Farm Cow Pasture, south-west of the main farm buildings, is entered from a lane that is a survival of the old medieval Milneway. It has a well preserved group of ridge and furrow with headlands and a broad balk. The lands are relatively high, and show a very marked aratral curve, cutting acutely into the headland. The field immediately to the west shows longer, narrower and much straighter ridge and furrow. We appear to have ridge and furrow from two different periods preserved side by side. The first group would appear to have been preserved by being turned down to pasture in the Middle Ages. In the adjoining field ridge and furrow, even if descended from medieval selions, has undergone the sort of modification which Dr. Kerridge described as common in the early modern period.[2] But very probably they are post-enclosure ridges made late in the Napoleonic Wars.

The measured acreages in the schedule confirm Maitland's impression that the acreages entered in the original, probably in the reign of Richard II, tend to be high.[3] For instance, in the seventeenth furlong of Middlefield there is a block of thirteen selions. These appear on the schedule of the Blackmore Map as thirteen ridges, containing 4A 1R 31P. In MS Add. 2601 they appear as four selions together with nine others adjacent, making a total of five acres. In the Corpus Christi "Hodilow" Farm Terrier of 1549 they appear as thirteen selions of five acres, but an eighteenth century hand has added the "correction", "6½ acr.". Similarly item 92 of the Blackmore Map schedule is given as 14 ridges containing 3A 2R 2P, but in the fourteenth century terrier these same fourteen selions are noted by the later scribe as five acres. The Hodilow Terrier merely repeats the larger area with no later addition. A sequence of three Corpus Christi strips affords a further illustration:

1 M. Beresford, 'Ridge and Furrow in The Open Fields', *Econ. H. R.*, 1948.
2 E. Kerridge, 'A Reconsideration of some Former Husbandry Practices', *Ag. H. R.*, 1955. For "aratral curve" see p. 54.   3 *T. & B.*, p. 128.

THE WEST FIELDS OF CAMBRIDGE

	Schedule 1789			Add. 2601		1549		18th cent.
Ref.	Ridges	Area		Selions	Area	Selions	Area	Area
No. 93	3	3R	30P	3	1A	3	1A	1½A
No. 94	2	1A 0R	16P	2	1A	2	1A	1¼A
No. 94	3	3R	20P	3	1½A	3	1A	1½A

This again suggests a very close correspondence between the modern ridges and the medieval selions, and confirms our earlier findings on the relation between the estimates or acreages of the terriers and the measured acres. In any case, one suspects that the Corpus estimates are based on little more than a counting of the heads of selions and adjusting by doubling for very long selions, and halving for short. Although the few acreages entered in the original hand do not belong to Corpus Christi College, and therefore have no equivalent in the schedule of 1789, the remaining acreages are still from the late Middle Ages, and can be directly compared with the modern measured ones. Moreover, marginal notes in the St. John's Terrier of 1617 give measured acreages for the Morris lands, and examples of the comparison are given below:

Morris lands in the St. John's Terrier of 1617[1].

Page	Selions	Estimated area	Measured area
6	1	2 rods	100 perches
7	2	2 rods	113 perches
9	8	4 acres	4 acres 2 rods 12 perches
21	1	1 rod	38 perches
21	8	3 acres 3 rods	3 acres 24 perches
22	6	2 acres 2 rods	2 acres 30 perches.
27	2	1 acre	108 perches.

Some totals of groups are given, and these illustrate the net effect of estimation and rounding to the nearest rod.

Page	Estimated	Measured
33	40 acres 3½ rods	38 acres 109 perches
49	86 acres 1 rod	65 acres 2 rods 13 perches

Headlands

Headlands are necessary concomitants of ploughing, since their primary purpose was obviously to enable the ploughteams to turn at the end of the furrow. In some cases this could be conveniently done onto a green way or road margin, but more often the ends of one set of selions abut on to other arable. For in an agricultural community where arable was precious and limited, headlands could not be left as waste, but fitted into the rest of the field lay-out in a tight interlocking jigsaw. In a few places, as at the side of Braderusshe or Willows Brook, a very few selions lying transversely to the adjacent furlong look as if

1 St. John's College muniments, xxxi. 29, 1617.

they must have been carved out of former pasture or waste serving as a headland. And there are a few cases where a headland is said to belong to an owner whose land neither flanks nor abuts it. This again could be a relic of more ancient practice, since most of these headlands belong to corporate owners, such as the Hospital or the Prior of Barnwell, where original benefactions were made at an early date. The Terrier always makes it clear to whom the furlong belongs where one piece of ground serves two purposes. Normally it belongs to the owner of the transverse strip in the next furlong, but where the owner of the turning plough cannot or may not make use of his neighbour's land in this way and the headlands are reckoned in with the selions of the same furlong, the words *cum forera* are used.

By what is probably not a coincidence, a transfer of meaning, similar to that noted for "dole", can also be found in North Cambridgeshire for the word "haveden". Etymologically just "heads", it was, by the early modern period applied to an access way, when the form had become obsolete in normal usage. In the West Fields, the compound name, "Clinthaveden", is used with enough imprecision of meaning to make it hard to be certain whether it refers to a piece of land, a boundary, or the Clint Way; and other similar difficulties occur over its use in the early charters of Madingley. But these shifts become comprehensible when considered in the context of the nature and variety of headlands.

There is consequently variety in the descriptions of the headlands. Most often a headland doubles as a selion lying across the lands' ends of the adjacent furlong. Sometimes only part of a selion is a headland in this way. Sometimes selions are said "to contain their own headlands". On other occasions a headland is the equivalent to a balk where a strip or group of strips lie between a headland and a balk. For example:

"4 selions of John de Coton formerly Thomas de Comberton's, the first and most east of this furlong, of which 1 selion is a headland. 3 selions of the aforesaid College, formerly of Thomas of Cambridge, next, of which the first selion is shorter than the other following selions at its north head, and the middle selion is in part a headland at the said head."

"4 selions of the Prior of Barnwell, the first and most northerly of this furlong, and the first selion is a headland." (fol. 11v.).

On occasion the word headland is used merely to indicate how the selion lies: i.e., immediately transversely where a broad balk lies against the heads of a group of selions, the selion parallel to and on the far side of the balk, is said to lie as a headland *(quasi forera)* to them. (fol. 17r., fol. 30v., fol. 19r.).

Headlands never seem to be common, except with the rest of the field, although there are rare instances of headlands in the later terriers being left unploughed. But it would seem that there could be doubt as to whether lands' ends could be common if there was no headland, for in the 1617 terrier, at the time of the great debate on waste and common, there is a note in the tenth furlong of Carmefield:

"Hitherto all the selions do abbut upon Bin Brook particularly, which argueth the lands ends not to be common". This may have been a very peculiar case where the teams turned on land too near the brook to be sown.

Strips and Doles:
The Units of Tenure

When Seebohm wrote of the half acre strips having lost their meaning he may not have been simply using the term strip as the equivalent of selion, but may have been thinking of a primitive state, albeit an imaginary one, in which the ploughland had been first allotted in units of a selion at a time. His discussion of open field units other than from the Cambridge evidence suggests this. In other words, he was not simply producing a confusion of terms, but interpreting by means of a model in which the two different units, that of tenure and that of ploughing, were originally identical. No one not already convinced of the applicability of such a model could find much support for it in the Corpus Terrier of the fourteenth century.

In that document strips range from the not uncommon single selion to the blocks of 45 and 41 (fol. 21v.), but where it has been possible to trace the provenance of individual strips through existing charters, it would appear that single selions are far more likely the result of sub-division by gift, inheritance and sub-holding rather than the basic unit. It is true that conversely families and institutions attempted to build up multiple holdings, but the names of the quondam owners and the differing tithe-owners reveal their composite nature. By the sixteenth century the drive towards the making of composite holdings was accelerated in those parts of the fields (usually the dampest or furthest away) which it was considered more profitable to put back to pasture and enclose than to keep as open arable. Such a piece is the one in Middlefield marked "A" in the Corpus Map of 1789. But the Terrier and the tithing still betray the fact that this is not an ancient unit. But such units are the exception, and fragmented arable strips the norm in the fourteenth century.

Some larger blocks there are, however, which tithe to a single recipient, and which, from their provenance, would appear to have descended to their fourteenth century owner intact from an earlier period in which, as the evidence of earlier charters seems to indicate, larger holdings were the norm. Such blocks are likely to be found in the hands of Merton ex Dunning; the Prior of Huntingdon ex de Troubelville; the Prior of Barnwell ex various benefactors before and in the earlier part of the thirteenth century; St. John's Hospital by much the same sort of donation; and finally Mortimer, whose enormous block of 41 selions in Middlefield is almost unique in the West Fields, although this owner has other notable blocks at the top end of Little Field and near the Castle. The Mortimer properties are believed to date from a royal grant made just before 1200, and divided almost immediately with the Hospital. It is not therefore unreasonable to suppose that a terrier of the West Fields made at the end of the reign of Henry II would have shown more of these larger blocks held mainly by notable feudal families. Had it existed, such a document would have delighted Maitland.

Maitland was on the look-out for traces of the older units, such as the virgate, in the pattern of the West Fields. Only in one place, even with antecedent documents, has it proved possible to track down a 30 acre piece in unbroken descent. This is the "L-shaped" block of the clerks of Merton in furlongs 19 and 21, which an early Merton document describes as, "In Campis versus Cotes triginta acre in duabus culturis quarum una cultura vocatur Godivesdole et alia vocatur Le Dale,"[1] and is probably that land, "formerly of Ailgar the Noble" described in similar terms in another Merton document.

Several of the larger blocks just referred to bear the name "dole". The great Mortimer block might well have carried the name also had it not been used for other blocks in the cispontine fields.[2] The block called "Carmedole" has a close association with Mortimer, and its origin is treated elsewhere. Besides these there are ten doles of varying size, but mostly large, in the West Fields. It is noticeable, when they are mapped, that they are all, topographically or agriculturally, marginal land. That is to say, they either lie on the far side of the fields, such as Mordole (Grithow furlong 15), Sherriffesdole (Middlefield furlong 10), Erlesdole (Middlefield furlong 18) and Goidesmedole (anciently Godivesdole) (Middlefield furlong 21); or they lie in wet places liable to waterlogging, Porthorsdole (Middlefield furlong 6), Brunneforthdole (Middlefield furlong 8), and that Dukdole cited by Maitland (Middlefield furlong 14), or in other unpromising places, such as Nakeddole in Brembilfurlong (Grithowfield furlong 7), and Peperdole (Grithow furlong 12) where gravel was dug on the side of Grithow. Finally there is the mysterious Priorsdole (Grithow furlong 6) lying rather awkwardly on the slope of the hill, which, from the fact that Grithowweye formerly ran across it, may be assumed to have remained pasture until a relatively late date.

All these, as marginal land in one sense or another, were probably later assarts from waste, and "dole" would then make sense as the portion dealt or allocated to one owner, the proper Anglo-Saxon sense of the word. This word has an incidental complication in that it is used in many North Cambridgeshire villages to mean a field-path, and is used indifferently for the path and for the furlong to which it gives access. For instance, it is clearly used for a field-path in the Cottenham by-law which forbids "the feeding (i.e. grazing) of balks, doles and fore-ends" until after the sheaves are carted.[3] In the Cambridge fields the term is never used for paths, only for blocks of arable. But it nevertheless suggests the kind of distribution of land for assart which we think we can perceive in the Cambridge West Fields.

Balks, great and small

When discussing balks, Seebohm cited evidence from the Corpus Terrier to show their use as divisions between strips. Orwin, in so far as he read the terrier, noted only major balks, and denied the existence of the minor ones which Seebohm had studied:

"These references, and all the others in which balks are mentioned, are evidently to parish boundaries or to common ways. Indeed, there could be

1 Merton deeds, M. 1606.
2 *T. & B.,* p. 179.
3 J. R. Ravensdale, *Liable to Floods,* p. 111.

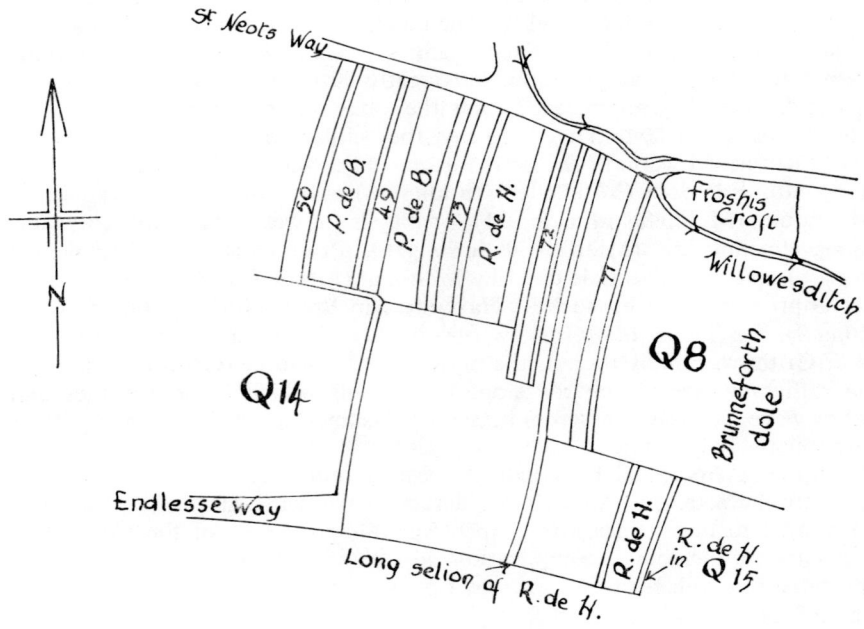

Nos. 49, 50, 71, 72, 73 are of C.C.C.

The long selion of Roger de Harleston in Q8 and Q14 and the piece at the end of Q14 which becomes merged with his land in Q15 adjacent.

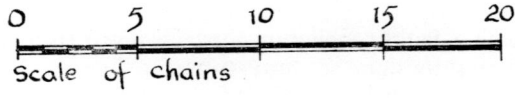

THE DESCRIPTION OF LAND

no conceivable reason for so clumsy and extravagant a means of strip demarcation."[1]

Certainly there are references to broad balks which were used as the Orwins indicate, and these were important functions. We find (fol. 3v.), "iuxta balcam que divisit campum de Cantebrig' a campo de Gyrton" at the far end of the fourth furlong, and again at the end of the fifteenth furlong (fol. 9v.), "ix seliones hospitalis predicti qui vocantur le Mordole iacentes in longitudine iuxta campum de girton quadam lata balca mediante". These broad balks which mark the parish boundary are only indicated where the ploughing in both the Cambridge field and that of the adjacent parish runs parallel: where one set of ploughing is transverse to the other no such form of demarcation is needed.

It is possible that these broad balks carried a fine stand of trees and shrubs, like the hedgerow which delights field botanists on the opposite side of the Huntingdon Road along the old Girton/Impington boundary. But the intense development of closes in the Howes area has made it difficult to disentangle the evidence of old hedgerow from new. Similarly the band of oak woodland enclosing furlong 24 towards Moor Barns has overlaid any traces of medieval broad balk boundaries. A sizeable balk, marking the division between arable and fallow, and broad enough to carry a fair stand of trees, is shown on Buck's *Prospect of Cambridge from the North West* of 1743.

The Cambridge-Coton boundary is of particular interest in that the broad balks have left their traces on the modern map, and in parts can still be walked, even if flattened out by modern farming. In the furlongs beyond the High Cross (Middlefield, furlongs 9 & 10), we find that there is no broad balk to furlong 9, because the Coton ploughing is transverse for almost all its length. But at the end of furlong 10 we find, (fol. 17v.), "unus selio Prioris de Bernewell.... et iuxta moram et campum de Cotes quadam lata balca mediante et est longior aliis selionibus sequentibus ad capud suum orientale." That is to say, the broad balk begins where the headlands of the Coton transverse field end, and from its edge the Prior of Barnwell has neatly acquired one selion (the acquisitiveness of the Prior of Barnwell in all the little edges, snippets and unconsidered trifles in this area is very noticeable). The broad balk and its companion selion make a little bump in the western boundary of the fields, which is clear on the map of 1803, and still shows on the Ordnance Survey editions of the 1930's.

Further down, where the old Coton Way and the modern Coton footpath cross the parish boundary, the broad balk dividing the parishes to the south was known as the Daleway. This path in its step-like course can still be walked, though it is now further marked by a ditch and a modern hedge. Shortly before coming to Edwin's Ditch at the far south western corner of the fields, the Daleway was met by a very important broad balk, referred to as the Weiebalk, which carried the Coton path through Dedale, and obviously preserved a common way which ante-dated the ploughing of Dedale, for it runs through furlongs 18, 19 and 20. The west end of the Weiebalk has been ploughed out, but its eastern half is marked by a ditch and hedge (though not such an interesting and ancient hedge as the one on the bank at the side of Edwin's Ditch). The Clintweye, a green way at the far south of the fields, may have had a balk.

1 *OF*, p. 45.

There is also a most interesting long selion with accompanying balk that may well be the remains of a former broad balk running from the St. Neot's Road through furlong 8 to the end of furlong 14, at which point Endlesseway formerly started. But by the sixteenth century another balk, slightly farther to the west, seems to have taken over its function and marked the division of the sowing seasons.

In the Carme Field two most striking balks, running parallel from north to south, dominate the map of the whole area. These are the Long Balk, on the line of the modern Grange Road, and Custes Balk. These served no function as boundaries in the medieval system and indeed cut right across several furlongs, demonstrating that they antedate the ploughing lay-out described in the terrier. If they ever served as boundaries, it must have been to a very small area of cultivation centred on Old Newnham. It has even been suggested that they originated in a pair of parallel dykes, such as are found in parts of Cambridgeshire (e.g. Fen Ditton), but flattened by usage for access and carting. But whatever their origin, they are one more of the peculiarities to be found in the Cambridge Fields.

But it is on the problem of whether, in spite of the Orwins' denial, there ever were minor balks as divisions between strips, that the Corpus Terrier and the other terriers of Cambridge West Fields can probably provide some of the best evidence that has yet come to light in this country.

In the fourteenth century Terrier, where a group of strips at the beginning or end of a furlong lie with no balks separating them, the scribe indicates this (almost, but not quite, invariably) by the marginal phrase *sine balca* and brackets. Occasionally he uses the phrase, *inter balcam et foreram*. Where such a group lies away from both the beginning and end of the furlong he writes, *inter duas balcas*. This implies balks dividing adjacent strips except where we are specifically informed to the contrary.

A calculation of the number of implied balks gives us the following totals in respective fields:

Grithow 226, Middle 388, Little 45, and Carme 149; a total of 808. The size of the total may disincline us to believe this interpretation even if we cannot deny the phrase *sine balca*, but collation with the later terriers makes it quite quite impossible to reject the implication. Where our earliest terrier uses the above form of negative notation to indicate the minor separating balks, the later ones use a positive form: *et b.*, or on occasion in full as *et balca*, or *balcatam ex utraque parte*, etc. The Clare Hall Terrier, in furlong 13, uses both systems: fourteen balks are noted, but after the final three strips, which have no balks, comes the phrase *et predicti quinque seliones iacent sine balcis*.[1]

Some of the balks changed during the life of the West Fields, and indeed they could be described as one of the least conservative features of the system, but only in contrast to the extreme conservatism of the rest. But whole sections of the later terriers, using positive instead of negative notation, depict a situation that fits the earlier picture exactly, and many show only slight change. This is not the result of slavish copying, for in the Clare Hall Terrier there are a number of corrections inserting balks that had been omitted. Further, some of

[1] Jesus College Terrars EST. 4.7.

the changes have been creations of balks, indicated by the phrase, *modo balca*.

In the St. John's Terrier of 1566 we get a new feature in two selions which have become balks as well:

iiij selions John Cotton, the first on the east part of this furlong, whereof the first selion is a balk and a headland. (This appears simply as *forera* in the fourteenth century).

j selion of William Alleine, now the first and eastward of this furlong, and is as a way.

In one section of the fields, however, balks seem almost to have disappeared by the sixteenth century, the furlongs of Grithowfield alongside the Huntingdon Road. These all lay on the highest land, where, for the most part, gravel lay over chalk, exactly the type of soil where we might expect separating balks to be most useful. But some balks at least appear to have come into use again in this area by the end of the eighteenth century: in the Blackmoor Map of 1789, which shows only the lands of Corpus Christi College, the College balks are clearly indicated and coloured green. A number of these appear in the furlongs alongside the Huntingdon Road in the area from which, according to the terriers, balks seem to have vanished earlier. Exchange of lands in this part of the fields makes the College balks here very difficult to identify with their fourteenth century predecessors, but some, at least, appear in the same places.

There is a further complication in the history of these minor balks: the form in which their presence is indicated in the earliest terrier merely notices their absence where there is a division between strips which lacks a balk. All the complete later terriers which we have studied assign balks to one or other of the adjacent strips, but never to both: they appear to have become more than a simple mark of division, and are shown to be narrow pieces of grassland appropriated to individual holders of the arable. This is very clearly brought out on the map, and the College balks can be counted. In Grithowfield there are 41, in Littlefield 30, in Middlefield 56, and in Carmefield 42, making a total of 169 balks belonging to Corpus Christi College. The preservation of the old terriers may have been in part the reason for the restoration of an earlier pattern.

Loggan's Views of the Cambridge Fields[1] were not in themselves satisfactory proof to a disbeliever of the existence of minor balks separating the strips in the open fields, but now that this can be conclusively proved from the documents, it seems clear on re-examining his prints, that the narrower strips shown (and these have no plough furrows) are meant to represent such balks. These illustrations, although very different from an exact measured plan, may possibly help us in trying to discover the function and purpose of these controversial features of the open fields.

There is an Ordinance of the Corporation in 1583 which defines part at least of their use:

It shall not be lawful to any person or persons to feed or pasture any horse mare or gelding, or other beasts *upon any balk or balks in the common field of the town, or to reap any balk until the corn of either side of such balk be*

---

1 D. Loggan, *Cantabrigia Illustrata*. Prospects of Cambridge from the East and from the West.

carried away or set in shock . . . . *PROVIDED, ALWAYS that this order shall not restrain any owner of any land, but his teamware may feed upon his balk next adjoining in the time of carriage of his corn from that land.*[1] (Our italics).

Professor Postan long ago suggested that where balks separated arable strips part of their purpose was to provide pasture. This certainly seems to have been so in Cambridge, and it makes very good sense. The Cambridge ploughs seem to have been particularly omnivorous, judging by the way they had taken almost all the land west of the river out to the borough boundaries. When Long Green was acquired by the riverside Colleges the last big grass common on that side of the river was lost to the commoners. The pressure on grassland must have been very intense in spite of the commons to the east of the river. However well the humans fared, the beasts of Cambridge must have existed very close to subsistence. Fodder had to be imported up the Cam from the fens.

The mention of reaping balks in the Ordinance must reflect this shortage of meadow ground: hay made from the weedy growth on balks that had been trampled by teams and men during harvest and cut so late must have been poor provender.

The Corpus Christi Terriers which deal only with the lands of their farms make rare mention of balks, but that of 1648 makes a very significant comment:

xiij selions next six selions of the Hospital balked 6a. 2r.

xiiij selions more next those 13 with a balk between them this balk between those 13 and 14 selions is *for the division of tithes.* (Our italics).

This may well have been the most important peculiarity of the town fields of Cambridge compared with the open fields of most Midland villages. Over the patchwork left by the plough lay the patchwork of land-holding, and over both lay the intricate patchwork of the ownership of the tithes. Unlike many village fields where there was often simple uniformity (all tithes paid to one parish church) the West Fields showed such complexity in their tithe payments that Maitland pointed out that the tithe owner changed every few yards.

Such a peculiarity may explain the function of the balks of division a little more. Where an occasional user went out on his annual trip into the remoter part of the fields to mark and collect the tithe sheaves, he would need guidance in the veritable tithe maze of Cambridge, and physical guides in the form of balks between strips, as well as the written record, might be virtually essential if ever the right sheaves, and only the right sheaves, were to be collected. But if the balk was to aid clearer distinction of tithe, it would incidentally give a clearer distinction than was normal of landholding, and this is confirmed by a comment in the Corpus Christi Farm Terrier of 1648:

This piece lieth in a wet furlong, and in times past by reason of the wet, part was left unploughed and next thereunto the Prior of Barnwell and three selions containing j acre which lay unploughed also, but now being all in tilth and not balked the Prior of Barnwell have taken away one land and so there remain but x, and his are made iiij, and thus it hath continued about some xxiiij years and not above.

In spite of the Orwins' finding their use as divisions inconceivable, those who held land in Cambridge West Fields were clearly liable to loss without them.

1 C. H. Cooper, *Annals of Cambridge,* 1842–1908, ii, p. 392.

THE DESCRIPTION OF LAND

There can be no doubt that, perhaps for peculiar reasons, the West Fields of Cambridge contained such minor balks for division between strips as a normal part of their arrangements. There were about a fifth as many of them as there were selions. They served particularly to make clear the divisions for tithe, and were probably used for access for the collection of sheaves. (This might not have not been quite so necessary for selions that abutted on the major roads.) They also seem to have been valued as a barrier against encroachment. We do not know how large they were, but their grass seems to have been worth something for feeding or mowing. Our fourteenth century terrier does not assign them to particular owners as do the later ones, but this may be simply because it was essentially a tithe book, and so limited to a description of the arable. Consequently we cannot necessarily infer that they were common at the earlier period. Later their use as common was certainly restricted, and they may have been common only with their appropriate field. If they originated as a convenience for tithe-owners and an inconvenience for the tillers, the tillers seem to have found ways of putting them to very good use in the end.

Furlongs

The furlongs, as we find them in the Cambridge West Fields, hardly look at first sight as if they could here have been the constitutive element of the open fields, as Finberg has suggested that they were in Devon and East Anglia.[1] Not only do they vary in shape as well as size, but also there is great difficulty in deciding what is a furlong. Opinion and usage in this matter clearly varied during the life of the fields, and some of the later terriers reckon them differently from the earlier ones thus creating difficulties in comparison because of differences in numbering.

The idea that a furlong is a bundle of selions which run alongside each other, thus forming a block, seems to be the general rule, but it is honoured almost as much in the breach as the observance. If followed literally it would produce many tiny furlongs that would complicate the system and be a nuisance to any scheme of reckoning. These small blocks of a few selions are normally lumped in with an adjacent larger one, even though it runs in a different direction; but, in order to help with identification, the terrier then goes on to explain that they form a furlong by themselves, or are "*quasi quarentena*". The result is that the first furlong of Grithowfield, for instance, as recorded in the earliest terrier, can be counted as one, two or three furlongs. The Sale Piece, which lay on the further side of the Huntingdon Road at the top of Castle Hill, was omitted from the Corpus Terrier and inserted in a later hand in the first furlong. The Clare Hall Terrier[2] counts it in the second furlong, and it forms a furlong by itself. In counting furlongs, Maitland arrived at the following figures, partly because he counted separately some of the groups that were normally included with a larger furlong: Grithowfield 30, Middlefield 24, Littlefield 11, and Carmefield 14 making a total of 79.[3] The scribe of the Corpus Terrier gave the numbers as

1   H. P. R. Finberg, "The Open Field in Devon", *West Country Historical Studies,* p. 146.
2   Jesus College EST. 4.2.
3   *T. & B.,* p. 124.

24, 23, 10 and 14 respectively, making a total of 71, and even then he was in error in his running total. The jigsaw-puzzle of furlong shapes, familiar from maps of the open fields, was thus especially complicated in Cambridge. It was made more complicated still in places where long selions ran through adjacent furlongs or where, on the other hand, short selions broke what might have been a straight boundary. Consequently we required all the supplementary documentation we could muster to complete the "restoration of our defaced open fields".

Study of the supporting documents makes it increasingly clear that the pattern of furlongs found in the fourteenth century was the result of a long process of piecemeal expansion of the arable. Assarts here and adjustments there, amalgamations and subdivisions, had utterly confused the outline of any basic scheme that might originally have existed. That this should have been so was almost inevitable, in view of the various invasions and extensions of settlement to which the Cambridge area had been subjected in the centuries preceeding the Domesday description. As we shall discuss in more detail in a later section,[1] in the pattern of the furlongs we think we can see traces of the stages of the expansion of the local community reflected in the development of their fields.

Fields and Seasons

As Maitland suspected, the system of nomenclature used for the great fields in the Corpus Terrier was comparatively recent, replacing the older system in use in and up to the thirteenth century.[2] By the time our terrier was written, these older names seem to have fallen into disuse. From then on, as long as they last, the names of the great fields in all the terriers which we have studied are consistent: Grithowfield (sometimes corrupted into Greathowfield), Middle Field, Little Field, and Carme or College Field. This is perhaps a little surprising in view of the way field names are used loosely in many Cambridgeshire villages, even where fields and shifts are coincident. In the documents which we have used, the only departure from this rule is in the Schedule to the 1789 Map of Blackmoor Field, which gives an additional two names, Little Carm Field and Dedole Field. These are obviously names of convenience for the easier location and identification of the strips numbered in the Schedule, and should act as a warning to those who would see a new assart or cropping shift in each new field name.

Maitland found much difficulty in calculating the acreage of the fields. He thought that he had missed about seventy acres because the size of some of the strips in Grithowfield was not given.[3] In fact all these acreages save one are recorded, but they are sometimes removed to the other margin, given as superscripts, or entangled with other notes. He suspected the terrier as describing full 1520 acres estimated, which would represent much less in terms of measured acres. Our count of the acreages recorded in the Corpus Terrier gives a rather smaller total.

1 Hints of Origins, p. 52.
2 *T. & B.*, p. 65. For a discussion of the older names *see* below on *Nomenclature*.
3 *T. & B.*, p. 127n.

THE DESCRIPTION OF LAND

	A	R
Grithowfield	509	2
Middle	604	1½
Little	113	
Carme	238	3
	1465	2½

The St. John's Terrier of 1617 gives the following totals:—

	A	R
Grithowfield	478	0½
Middle	642	1½
Little	122	2½
Carme	240	1
Total	1483	1½

When allowance is made for the few strips for which no acreage is recorded in the Corpus Terrier, the two totals would be even closer.

Each of the great fields formed a block surrounded by a continuous boundary. While these need not necessarily have coincided with the seasons, there is a fair probability that at one time they did so. Indeed it seems likely that the small running numbers in the lower right hand margin sof each folio of the Terrier indicate the seasons as they were cropped in the fourteenth century. The numbers run 1a, 2a, 3a, etc., throughout Grithow Field, 1b, 2b, 3b, etc., throughout Middle Field, and 1c, 2c, 3c, throughout Little Field with Carme Field.

By the early seventeenth century some enclosures had taken place, chiefly from Grithow and Carmefield, so that Middlefield was even more disproportionately large than it had been in the fourteenth century. At some point the seasons had been adjusted accordingly. The St. John's Terrier of 1617 breaks the sequence of the fields in order to re-arrange the material in seasons. It begins:

> A perfect Terrar made out of Blackamore, of all the lands lying in the fields of Cambridge in the West, divided into three several seasons or times of ploughing. For one season or part of the field lieth fallow every third year, and so they are all set down in this Terrar as they are severally and yearly laid out.

The first season was divided into two parts, "the one distant from the other half a mile at the least, yet they are ploughed and sown together yearly, and alter not." There does in fact seem to have been some slight alteration at the margins, but this was insignificant. The first part of the first season was made up of a block in the south-east and south, Carme Field (or College Field), Littlefield, and three adjacent furlongs of Middlefield, the 21st, 22nd and 23rd. The second part of this first season was the furthest part of Middlefield, west of a line roughly at right-angles across the head of Endless Way, cutting through the eighth, fourteenth and fifteenth furlongs.

The second season consisted of all the rest of the Middlefield not already mentioned as attached to the first season, except for its first furlong.

The third season consisted of all Grithowfield and the first furlong of Middlefield, a simple topographical unit between the Madingley and Huntingdon Roads. Thus in spite of all its peculiarities, this field system seems to have achieved something like orthodoxy in its rotation.

The size of the three seasons is given in 1617 as:

	A	R
First Season	572	1
Second Season	393	3
Third Season	517	1
Total	1483	1

Although unequal, the seasons were much less so than the fields. They could form effective units in a three course rotation, and it was for this purpose that they were designed.

We would have welcomed more detailed evidence of cropping, but had to be content with the two account rolls of the Priory of St. Radegund for 1449–50 and 1450–51. Even there a complete picture could not be obtained, since the entries on dorse of each roll (not printed by Gray), proved to be so badly rubbed and stained that fragments only could be made out. This was a particular disappointment since they included part of the return of grain.[1]

The third roll, for 1481–2, has much less information, and this suggests a house in deep decline and much less active as a cultivator. The earlier rolls show the Priory cultivating most of its demesnes in Cambridge, although it was leasing its village lands and farming its tithes. No location of its crops in the fields is given other than a mention that its tares were sown in Bartonescroft. We are unable, from such evidence, to observe the operation of the seasons of the West Fields, or even to diffentiate its crops from those of other Cambridge Fields. We would very much like to discover account rolls detailed enough to enable us to do this.

Such information as we have, however, does enable us to see a little of how Cambridge fields could be involved in the economy of a local religious house. The main field crops seem to have been wheat and barley, but seed was also purchased for oats, various kinds of peas, tares and mustard. Much of the produce was sold at the market, but some privately. The wages account shows the normal agricultural tasks being performed by men and women employed from outside for each specific piece of work. In the same way some of the wool from the demesnes was used for the household, but spinning, warping, weaving, fulling and shearing were performed by outsiders on separate contracts. Very significant, in view of the shortage of pasture in Cambridge, is a large item of expenditure for the agistment of beasts in Willingham with its vast fens.

Enclosures

Cambridge, in its growth, had clearly overreached the capacity of its fields to sustain it by the time for which we have detailed record evidence. The availability

[1] A. Gray, *The Priory of St. Radegund,* CAS Octavo, xxxi, Cambridge (1898).

of grass in the fens meant that expansion was felt most in an unrelenting pressure to maximise the arable. As far as the West Fields are concerned this is shown very graphically by the First Draft Map for the enclosure of St. Giles' Parish. The areas there coloured green are very sparse. Maitland commented on the lack of enclosure that had taken place before the open fields were finally swept away in the nineteenth century.

When we come to plot all the arable shown in the Corpus Terrier, the absence of green is even more striking in the fourteenth century.

From his studies in the history of land-ownership alone, Maitland was convinced that the system of the Corpus Terrier, which remained so stable thereafter, could not have been found complete at a date very much earlier. At the time of the survey which must have preceded the writing of the Barnwell Tithe Books, the fields had expanded to their uttermost limits, in some parts we might say beyond their limits. Marginal lands left unploughed earlier had been absorbed into the arable and there was virtually nothing left to plough that was not gravel-pit, claypit, or liable to flood. By the time of the writing of the Corpus Terrier, there are signs that the tide had turned and the arable had begun to retreat.

Considered topographically, the first retraction took place chiefly on the far Western boundary. The Corpus map of 1789 omits certain furlongs altogether in this area, though they had contained strips belonging to the College in the fourteenth century. It would appear that Cambridge owners, who had been active enough at the time the last assarts were made from the moor to have pushed out the Western boundary, lost interest in the retention of far-flung strips earlier than their neighbours in Coton and Madingley, who seem to have taken up any 'spare' land, even if of doubtful quality. By the early sixteenth century, as noted in the St. Johns' Terrier, the last furlong towards Moorbarns had been acquired by owners based in Madingley, and was enclosed, and virtually incorporated into the Madingley estate of the Hynde-Cotton family. By a little chicanery, the ambitious Aunger family managed to take over most of the former Corpus holdings in the furlongs beyond High Cross by about 1580, and add them to their growing Coton possessions, no doubt with an eye to keeping up with Hynde's.

On the Girton boundary, consolidation of strips, destined to produce the larger Howes close we find on the 1789 map, had been achieved by the mid-fourteenth century by Thomas Atchurch of Howes. There were already fewer small tenants here than are mentioned in Merton charters of the previous century. Spalding's close, the former furlong 14, was enclosed and given up to gravel pits by 1500. But medieval digging was shallow and by 1617 some of the worked area was partially restored.

> "1 selion of Thomas Atchurch of Howes, the first and most south; Blackamore saith it is an headland, but it is now altered and is not so by reason much gravel hath been digged and made pits, and now ploughed again otherwise than it hath been at the beginning."[1]

Some enclosure from quite early times seems to have come from the needs of the town for gravel. The main building complex of the University farm sits

---
1 St. John's College. xxxi. 29, p. 43.

among the irregularities of the medieval pits of the Grithow, centred on a point known in the fourteenth century as the Greenplat. In the Corpus Terrier, selions of the fifth furlong are shown as abutting on the Greenplat, but the mid-sixteenth century hand has noted against it, 'gravel pits'. The pits engulfed it and spread, to the confusion of the arable here:

> "The furlong in which is Grithow Hill lying on the farther side at the north head of v butts of the land of Merton College and abbut upon Grithow Path. The selions of this furlong ought to be counted at their east head but yet it is better to begin now contrary next the conduit for it is now in the gravel pits and can scarcely be descried where the beginning of this furlong appeareth".[1]

There is a Latin version of this in the Clare Hall Terrier, dated 1538. This became the site of the Observatory. When the shallow gravel deposit had been exhausted here, the area became rough pasture closes, of which the pitted surfaces are still visible.

Clay was also being dug in some parts of the fields.

> "Note that these 4 last single selions are now made claypits, and not ploughed."[2]

But it was more usual for clay digging to take place in ground too water-logged to be suitable for arable, for instance at the end of Smalemade, where early desultory digging was followed after 1800 by the creation of the extensive pits that are still extant (off Clarkson Road). The main clay pits for the upper town were at Asshwykston, by Castle End, and may have begun as gravelpits.

Apart from the relatively slight contraction of the limits of the West Fields between the fourteenth and nineteenth centuries, there is evidence for some reduction in the area under arable cultivation within the West Fields themselves during the later Middle Ages. The first pieces to go out of cultivation would seem to have been the odd butts and transverse strips along road and stream margins.

> "Note there are two selions which do lie cross of part of this furlong of the monastery of Denny and next along to Willow Ditch, which are not ploughed, and they seem to be a little furlong by themselves."[3]

Adjacent to these the monastery had five selions in the next furlong which it left unploughed with them. These, however, did not necessarily become enclosed. In the chronically wet lands of the low-lying clay in the far West of Middle Field we can see consolidation and enclosure working together to produce small pasture closes, from the St. John's Terrier.

> "11 selions of Benet College, whereof 4 are longer than the former at the west end and they are now divided from the other 7 by a ditch set with whitethorn.
>
> 4 of the Prior of Barnwell ... Note that these 4 lands are in possession of Benet College by exchange for certain lands inclosed in Moor Barns Close."[4]

Next again, Dukdole was enclosed by the Prior of Barnwell by ditch and hedge,

1 St. John's College, xxxi. 24, 1566.
2 St. John's College xxxi. 29, p. 40.
3 *Ibid.* p. 23.
4 *Ibid.* p. 17.

and a large T-shaped enclosure was made by Corpus Christi College out of a group of selions and butts. At some unspecified date, Merton College made a close in the corner formed by a bend of Edwin's ditch in Deddale. In Grithow Field, the Morice family allowed a block of land next Seman's Ditch to revert to pasture in the late Middle Ages, and here the form of the selions has remained undisturbed to this day. Below the Grithow, Froshiscroft, also called Freshcroft, had become Hunnell's Croft or Close by the sixteenth century.

Here and there, other small patches surrounded by hedges appear on the 1789 map, but we have no means of discovering when the hedges were made, or whether the enclosed ground was at once turned over to permanent pasture. In some cases it is clear that an area under single ownership, surrounded by hedge and ditch, remained under arable cultivation as late as the end of the sixteenth century. This is the case with Muscroft, while the Sale Piece described as a croft in the Terrier, is shown being ploughed on Hammond's map of 1592.

A *caveat* seems appropriate here about the terminology of the Corpus Terrier with respect to the terms "croft" and "close", as shown by supporting documents. The former term, "croft", denotes a compact holding under single ownership, often identified by a personal place-name element and sometimes also connected with a messuage bearing the same name. Some of these are of great antiquity, others may be of more recent formation. The significant factor. according to the terrier of 1566, is that they may be "every year land at the will of the owner", that is to say, they need not run with the seasonal shift followed by the rest of the Field. In some cases this may mean more intensive cropping, as the writer of the explanation implies: in other cases it can mean withdrawal from arable cultivation. Some fencing, if only a ditch and some stakes for hurdles, would be necessary to mark off a croft from the neighbouring land at those times when it was not conforming to the rest of the Field. It is therefore not surprising to find that the crofts of the thirteenth century and fourteenth century charters and terriers tend either to merge in the general arable, if they remain in arable cultivation, as we find is the case with Cupitts Crofts,[1] or to be permanently withdrawn from the cycle and surrounded by a more permanent hedge or fence. Then they become what are called "closes", or as the Terrier says, "*nunc in clausuris*".

Newnham Crofts are of particular interest. The ancient crofts were on the good arable land of the Barnwell terrace gravels, and had been absorbed into the general arable, bar the name, by the time of the Corpus Terrier. These remained for the most part in arable cultivation to the end of the life of the West Fields. It is only towards the late eighteenth century that hedges creep round the ends of the selions abutting on Old Newnham Way. These could have been planted as some sort of protection against the increased numbers of men, beasts and vehicles using the turnpike road, and the verges were considerably widened at Parliamentary enclosure. One pasture close certainly created before 1580, the close in Newnham belonging to Clare Hall, seems to have been made because of its proximity to the houses, and it is here that we find the first new buildings

1 From hence do follow Cupids or Cupis Crofts, which are thought to be every year land, but of late they have been added to the second part of the first season. . .

encroaching upon the arable. One such house is found in the personal possession of the Master of Clare College in the late seventeenth century.

From the beginning of the seventeenth century there were signs that the desperate need of the town for building land might have been too strong for encroachment on the fields to be avoided:

> This part of the furlong beginneth at 2 selions of Benet College on which 2 selions ther was erected a dwelling house in anno 1603.

and again:

> Next unto this Castle Lane or Barton Way there is another close containing 4 selions, sometime Richard Tuliett, and 3 selions sometime Thomas Boll, in which close there is a tenement built viz., in anno 1607. It is also called St. John's Close.[1]

In all these cases we have discovered only collegiate owners are involved, and they build on small enclosures they have previously made from portions of their lands adjacent to existing dwellings. All in all, what is surprising is not that some houses went up on the fields but how few in number they were before 1800.

On the Eastern edge of the Fields further retraction took place from the sixteenth century under the double pressure produced by competition for common pasture and places for recreation in the riverside meadows. The Corpus Christi Farm Terrier of 1648 shows:

> "(. . .land of) Roger Harleston which is now green sward and is made common but belongeth to St. John College. This furlong doth abbut upon Bin Brook and Long Green near unto Garret Hostel and Trinity College Close, six selions lying along a broad balk being an usual and common place to walk in. . .ix selions on the back side of St. John's College abbutting upon Bin Brook towards the west. These be enclosed by St. John's College by consent."

With the aid of the terriers we can trace something of the creation of the Backs. (See below, p. 80).

1 St. John's College. xxxi, 29, p. 16 and p. 32.

THE MAP OF THE WEST FIELDS

The Area of the West Fields-Boundaries, Routes and Crosses

The Cambridge West Fields are possibly the most neglected area of study in the whole of the County of Cambridgeshire over the last decades. This has come about primarily because the major works published in series relating to the counties of England, notably the Victoria County Histories and the volumes of the Historical Monuments Commission, work to a scheme based on parishes for the country districts and on towns for the urban areas. The Cambridge West Fields fit into neither category, technically belonging to one of the Cambridge city parishes and yet being too rural for the urban historian to treat more than cursorily until very modern times, when the rapidly expanding town pushes out into the former rural area, destroying as it does so the very features the historian of the agrarian scene would most wish to record. Other writers, dealing with a variety of aspects of the history and development of the Cambridgeshire, tend to fall silent as soon as the boundaries of the Cambridge parishes are reached, and their excellent and informative maps of routes, settlements or archaeological finds only too often show a blank starfish round Cambridge itself. Conversely, accounts of the growth and development of the town are accompanied by maps and diagrams which stop up to two miles short of the Cambridge boundary, though this is preferable to the practice prevailing at the beginning of this century of writing medieval names into a modern draft outline map.

It is to be hoped that when what we have been trying to do for the West Fields has been done for the East Fields (the pioneer work of H. P. Stokes has shown that the documentary evidence is ample), and ideally also for the Chesterton Fields, it will be possible both to map and to describe early, medieval and near-modern Cambridge in its rural setting more fully and accurately than has hitherto been customary.

And it is not our knowledge of Cambridge only which should profit from further study of its immediate environs. One unfortunate result of studies based on a parish by parish description (usually arranged alphabetically, though sometimes Hundreds are, rightly, preferred) is the difficulty of dealing adequately with those communities which never achieved parochial status, or which achieved it late, or which sit astride the parish boundaries of the more important centres, and so tend always to be of peripheral interest to the main subject of the section. The Cambridge West Fields contain parts of two such settlements on its extreme boundaries, Howes, to the North East, and Newnham to the South. Newnham is particularly intriguing, in that it is itself a village in two parts. The main part, near the mill, has an ambivalent relationship with the town across the river and its built-up area is divided between two of the cispontine parishes. The other part, Newnham Crofts, has an ambivalent relationship with

Grantchester, and the crofts themselves are divided between the two parishes. It is clear that neither the early history of Newnham, nor indeed the full history of Grantchester, can be written without a study of their surviving archive evidence from all four parishes.

Even more intriguing is the problem of the early development of the Cotes and its ascent to manorial and parochial status. Anyone who has worked on the early archive material of the Barton/Whitwell and Grantchester/Coton area will be only too well aware that here again a parish by parish plan of study is sadly inadequate to deal with the complexities of the situation which existed here at Domesday and for a couple of centuries afterwards. By the later Middle Ages, the efforts of the greater landowners, mainly the Colleges, to rationalise conflicting patterns of feudal landholding, traditional agriculture and ecclesiastical parishes, resulted in some places in a superficial approximation to the ideal of "one village, one manor, one church", which had never obtained in former times. And the ultimate success of St. Catharine's College in Coton has perhaps blinded historians of the sixteenth and later centuries to the thinness of the paper over the cracks. Yet it seems fairly certain that the Cotes were once in an ambivalent position on the fringe of the West Fields, similar in many ways to that of Newnham Crofts or Howes Crofts. Topography, trackways and analysis of ownership point to a situation in which the stronger Cambridge-based community at once threatened to overleap its bounds to create and appropriate agricultural land on the further side of the ancient boundary, and at the same time seems to have been something of a magnet attracting population from declining Whitwell to expanding Coton. Charter evidence from the late 12th c. suggests that the Burdeleis family, Lords of Madingley, claimed feudal jurisdiction in the area. At the same time Coton was technically a part of Grantchester parish, a village to which there was not even a direct road from Coton. Clearly, the history of Coton cannot be written without reference to what was taking place in the West Fields of Cambridge, and without using the fairly numerous charters which provide evidence of extraneous ownership in those parts of the West Fields nearest to Coton.

Even the Huntingdon road, which at first sight would seem a logical, almost inevitable, boundary, turns out on investigation of the charter evidence unlikely to have been used as such in the early period of settlement. Although this ancient Roman road serves as a boundary for parishes west of Girton, Girton parish and field system cross the road into the valley of the Beck Brook. There are reasons for believing that the inhabitants of Bede's *Grantacaestir* likewise farmed an area on either side of the Road and are more likely to have used Braderush and Willows Ditch, or Edwin's Ditch and the Binn Brook, to mark the limits of their territory. Following the Cambridge boundary in a circuit from the Huntingdon Road at its crest in Howes, we note a "stepped" boundary, such as is usually thought to be indicative of the presence of arable antedating the fixing of the parish boundary. But beyond the gravel ridge on which the Howes stands, from the stream running out from Kynchmade, the straight boundary suggests a stretch of former uncultivated pasture, continuous with Madingley Moor. Had this boundary been continued on its same alignment it would have crossed the St. Neot's or Madingley Road at the point where the former High Cross once stood, as described in the Terrier. Two furlongs further south the boundary again falls into the same alignment. We think that once the boundary must in fact have run like this, but was pushed further to the West to take in assarts at Moorbarns and the "furlongs beyond High Cross".

So once again, from the Madingley Road onwards, there is a "stepped" boundary, this time with Coton. But Coton parish is a late creation (c. 1381) and at the time of the fixing of the boundary the adjacent parish was Grantchester, though one cannot imagine that many of the actual inhabitants of Grantchester village were involved. The "stepped" boundary continues into Dedale, which from charter evidence seems to have been ploughed up by the twelfth century though perhaps not long ploughed at this date.

But the southern boundaries of the Fields are different from the western ones in bearing no signs of carefully worked-out partition between existing agricultural communities. They follow the lines of two topographical features which must have been there before the arrival of the Grantasaete and other "Anglo-Saxon" settlers. The first is the stream known as Edwin's Ditch, which enters the West Fields from Coton and serves as boundary until it swings N.E. towards the Binn Brook. This brook was probably deepened and perhaps straightened after being chosen as a boundary and so came to bear the name of the unknown Edwin. The second is the very ancient trackway known as the Clint Way, named from the Clint or hill, which also gave its name to the Coton Clint Field, though on the 1789 map it is also called the Mare Way (mare=boundary). This way runs eastwards from the higher ground above the valley of the Binn Brook to its junction with the Fulbrook at what the Terrier calls Portebrigg. The bends in the brook were drastically rechannelled when the turnpiked road was made, and it is not easy to reconstruct its former course, but it would seem that the Clint Way originally crossed both brooks separately, just before they joined, and then carried on towards the river on the line later used by the turnpike, or modern Barton Road. On the Corpus Terrier this stretch is considered as part of Old Newenham Way. It probably ended at a river crossing near the present bathing-place on Sheep's Green, for in Barton parish documents it is referred to as Hinton Way. The last hundred yards of its course follow a small stream later known as Vicar's Brook.

If Cambridge and its immediate environs are studied on a map, and Chesterton is included with Cambridge East and West Fields, the whole appears roughly in the form of a wheel, with its hub on Castle Hill, its rim at the outer edges of the Fields, and with ancient routes into the centre as its spokes. Three such routes cross the West Fields.

The first is the best known, the Huntingdon or Godmanchester way, straight as Roman roads should be, and with a continuous history of use from Imperial times to the present. The second is also very ancient and probably pre-Roman, the track-way from the west that comes into Cambridge from the top of Madingley Hill. From Madingley westwards it also serves as a parish boundary. Its course meanders more than that of a Roman road and even widening after turnpiking and enclosure have not served to iron out all its twists, though modern improvements are doing their best. From a point near Churchill College its alignment is almost directly towards Castle Hill, and it seems probable that its original course was up the Stoupendecruche Way straight into the town. The third route appears on the map as the Barton Way and is the last stretch of the Roman road from the South West, well authenticated as it passes through Barton parish and Haggis farm, but subsequently marked on maps by a "notional line" or declared lost.[1] Rising steeply after crossing Edwin's Ditch, it passed close to the back of St. John's Grange Farm and from this ridge was re-aligned onto the site of the Castle mound, turning about 10° East. Beyond the stream crossing at the end of Smale Made, where several green ways met, the medieval Barton Way again turned somewhat more to the west presumably to avoid the embankment remaining round the western edge of the upper town, the "fossatum regis". Here it was joined by the other medieval routes from the west entering the upper town at Asshwykston.

Through ways from North to South were also provided by the two very ancient long balks which ran parallel across the Carme Field, predating the ploughing which spread on either side of them by the Middle Ages. The more westerly of these, Custes Balk, had an extension across the Binn Brook to link it with the Barton Way junction at Smale Made. The other, Long Balk, did not continue across the Binn Brook until the creation of the modern Grange Road along its route after Enclosure. Both these balks crossed the Southern boundary of the West Fields to link with routes from Grantchester and it is possible that the balks themselves, if they represent the remains of early earthworks or dykes, may have continued across into the Grantchester Fields. The more used route was that opposite Custes Balk, roughly the present Grantchester Road. The line of Custes Balk is still marked by the avenue of trees which leaves the Barton Road almost opposite Grantchester Road and continues into the garden of Leckhampton House. It is shown in maps as a stub of a way as late as 1865,[2] but by then it gave access only to a piece of field. In the Middle Ages, as today, there was an alternative route to Grantchester by the Meadows.

With this basic network of ancient trackways, only a few more "green" ways were needed to enable the inhabitants of Cambridge or Newnham to reach any of the neighbouring villages. It is perhaps significant that no fewer than three such ways led from Coton, or two and a half if Endless Way is disqualified as only going half-way. The main Cotes Way branched off the Barton Way at Smale Made and it is from a few yards up this way that the famous Loggan print of Cambridge from the West is drawn. The second through way was the Coton Path, a continuation of Froshlake Way (from Newnham Mill) which ran beside Edwin's Ditch for part of its length and was thence carried on a Way balk to the Coton Boundary.

In the Grithow Field the place to which access was most needed was by the How itself, where gravel was dug. Grithow Way provided the most direct route from the upper town, Grithow path from the Madingley Road. For the inhabitants of Howes or of Girton, the Millway from the back of Howes Crofts led to the same point.

Crosses stood at many of the strategic points where routes and boundaries met or needed to be marked. Two crosses stood at the entrance to the upper

---

1   It appears, however, clearly on aerial photographs.
2   C.C.C. *XIV* 165.

town itself, the High Cross shown on the sixteenth century maps and the remains of its predecessor, Ashwykston or the Stump Cross, a little to the South. The Terrier says that Ashwykston was "a quoit's throw to the S.W." but this does not allow for the direction of the Huntingdon Road being reckoned as due North.

There were also crosses in the Madingley Road. The High Cross at the old western boundary is shown in the margin of the Terrier and was presumably still standing at the time of writing. By 1617 the writer of the terrier of St. John's College notes,

> "There is mention in Blackamore of a cross and a little hill standing in St. Neot's Way at the end of these furlonds there is none such now, neither hath been of late that can be remembered".

The stooping or bent cross, as shown in the margin of the Terrier, stood where two ways originally parted at the approach to the town. The first and possibly older route, named Stoupendecruche Way after the cross, had, by the time of the Corpus Terrier, lost the last part of its course to ploughing in Furlong 6 and had been turned down to the road by a balk bordering Muscroft furlong. After the stooping cross had fallen and been forgotten, the road was sometimes erroneously called Stop-end-cruche Way; as in the 1617 terrier, and by the 18th century also Stoneponde from the Stone pound on what came to be Pound Green. The second road, still called St. Neot's Way, continued towards the river, linking with Merton Hall, and possibly with a footway to the Great Bridge. But it was quite narrow as it passed the end of Little Field by St. John's Barns, and did not become a through way either into the town or across the Bin Brook to the Long Green until the sixteenth century. Thereafter the increasing use of wheeled vehicles and the consequent desire to avoid steep gradients made it the preferred main route out from Cambridge to the West and it was considerably widened at turnpiking.

By the sixteenth century, wooden crosses stood at several other points and their position is entered in the terrier in marginal notes. One such, Hunnell's Cross, stood at the Barton Way—Sheepcote Way junction, and another, Colys Cross, near the Clint Way–Barton Way junction. The 1566 terrier, describing Clint Way, says, "and it endeth at a cross of tree that standeth in Barton Way by a great row of willows that grow beside Portbridge". There seems also to have been a cross at the angled bend of Old Newnham Way.

Topography and Lay-out

At first sight, the division of the West Field as set out in the Corpus Terrier, which lasted with practically no major changes up to Enclosure in 1805, seems so obvious and logical, that we might well imagine that it had existed from time immemorial. In this system, the three most ancient trackways which traverse the area serve to divide it into three roughly equal segments; the first being Grithow Field, the second Middle Field, and the third Little with Carme Field, the division between these two last being the Binn Brook.

With two minor exceptions Grithowfield seems to be a logical topographical unit between the two pre-existing trackways that we know today as the

Huntingdon and Madingley Roads. The first minor exception is the Sale Piece by the Castle, on the far side of Huntingdon Road. It was forgotten in the writing of our first terrier, but inserted later on as an addition to the first furlong. At that time, its nine selions, two butts and a gore, were all in croft. The Corpus manuscript, like some of the later versions, is followed by two summary terriers of lands in Chesterton Fields, added in the fifteenth century in the blank folios (see page 9). There is considerable early charter evidence of early ownership in both sets of fields, particularly in Corpus, Clare and St. John's Colleges. Even as late as the first year of the reign of Henry VIII, a certain Margaret Rawlins bequeathed her family property, consisting of a farmhouse on Castle Hill and farmlands divided between Cambridge and Chesterton Fields;[1] old habits died hard.

The two sets of fields possibly look back to a common origin and subsequent division, never entirely satisfactory, since so many townsmen held land in both. In our opinion, the point about the Sale Piece is not that it intruded into the Chesterton Fields, as might appear at first sight, but that it was excluded from the arrangement by which the Castle and its precincts were deemed by the Norman conquerors, for reasons of defence and administration, to be part of the royal manor of Chesterton. And looking at the extraordinary pattern of the Hundreds on the far side of the Huntingdon Road, with Chesterton Hundred in three detached pieces alternating with parts of Northstowe Hundred, we may well suspect that the whole area here has undergone considerable revision in the arrangement of the Hundreds, again probably in connection with arrangements for defence in the late Saxon period. Too much, in fact, has happened for the boundaries we find on the post-medieval map to give any clues to earlier patterns of agriculture or settlement.

The second minor exception is the first furlong of Middlefield, Muscroftfurlong, separated from the rest of the field by the St. Neot's Way (Madingley Road). But the older route, as has already been indicated, appears to be represented by Stoupendecruchway. So it is consistent that this is the field boundary.

Apart from this first furlong, Middlefield made a simple topographical unit, with St. Neot's Way and Barton Way as its natural boundaries. Littlefield formed a similar natural unit between Barton Way and Binn Brook, and Carmefield another from the Binn Brook to the meadows and low lands by the Cam. It is also noticeable that while the fields reached from the Upper Town are roughly on a segmental pattern, those of the Carmefield are more on a grid.

A closer look at the topography and lay-out of the fields, and a consideration of older names found in charters for the fields and their parts, suggest a process by which an intricate pattern of smaller units was built up into the system found in the Corpus Terrier, analagous in fact to the way in which we believe the fields themselves to have been built up. Assarting and the expansion of the fields must clearly have taken a very long time, and there are hints, in the lay-out, of distinct phases. For instance, the westermost parts of both Grithow and Middle Fields along the Madingley Road are a complex of small furlongs that suggest late piecemeal assarts into the moors, and it is in this area that the frontier between corn and grass ebbed and flowed most clearly during the period

1 G. & C.C. *XVII* 1.

for which we have records. The area near the Coton boundary, (from the eighteenth to the twentieth furlong of Middlefield) which includes the part known as Deddale, or The Low, again suggests assarting from the shape and arrangement of the small furlongs, as well as from the "stepped" boundary already referred to. But it would be most unwise to assume from the map alone that the presence of a small regular-shaped furlong at some distance from a main centre of habitation is sufficient alone to indicate assarting. To what appears on the map we must add what can be discovered of the soil and conditions of drainage, of the presence of springs or unhelpful contours, otherwise we are likely to fall into misconceptions.

For this reason we have tried to walk over as much of the West Fields as possible and are most grateful to the Superintendent of the University Farm, who allowed us to explore the property, and patiently answered what must have seemed very odd and stupid questions. And there are various other farmers, gardeners and workers, whose names we never discovered but for whose helpfulness under interrogation we are grateful. Where the open land in question has disappeared under modern suburb, a descriptive field-name can sometimes supplement geographical data.

To illustrate the general principles just set out, we may compare furlong eight in Grithowfield with furlong eight in Carmefield, both small, square-cornered furlongs surrounded by larger irregular ones. Do they precede or are they subsequent to these surrounding furlongs? Grithow 8, the *parva quarentena*, has a field-name, "Dukmere", for a piece at the side of the western selion, and the modern farmer confirms that this is a place where he still has drainage problems, since here the gravel of the Howes ridge gives place to clay, and the surface is too level to drain off naturally. So on balance we decide that the *parva quarentena* is marginal land, liable to be waterlogged, and was probably left till last in the ploughing-up of the area, only being ploughed when there was great economic pressure to plough up everything that yielded anything.

Conversely, Carmefield 8 is on good, well-drained soil, on the gravel ridge between Custes balk and Long Balk, and from the way in which the adjacent furlongs give place to it, it may well be the earliest choice of land for ploughing up in this area. Intricate patterns of furlong-shapes are always a puzzle, but unless a very abrupt contour gives a reason for a change in ploughing-direction, it is fairly safe to say that the squarish or regular-shaped furlong usually precedes the oddly-shaped one that fits into the remaining spaces. An example of such an oddly-shaped furlong, which as it goes along incorporates small blocks of earlier cultivation, is furlong 6 of Grithowfield.

Another fairly general rule would seem to be that smallish furlongs close to ancient centres of settlement probably represent early small fields of small communities. Such fields are to be suspected in furlongs 1–4 (first part) of Grithow, and again in furlongs 1, 2, 6 and 13 (south part) of Newnham.

Long sweeps of furlongs out towards the western boundary, on the other hand, probably represent large-scale ploughing-up by a larger community. The huge furlongs (15–17) of Middlefield are the best example of this type and since these are in fact on heavy clay soil, and would need a fairly large output in man- and oxen-power to bring them into cultivation, are almost certainly a result of

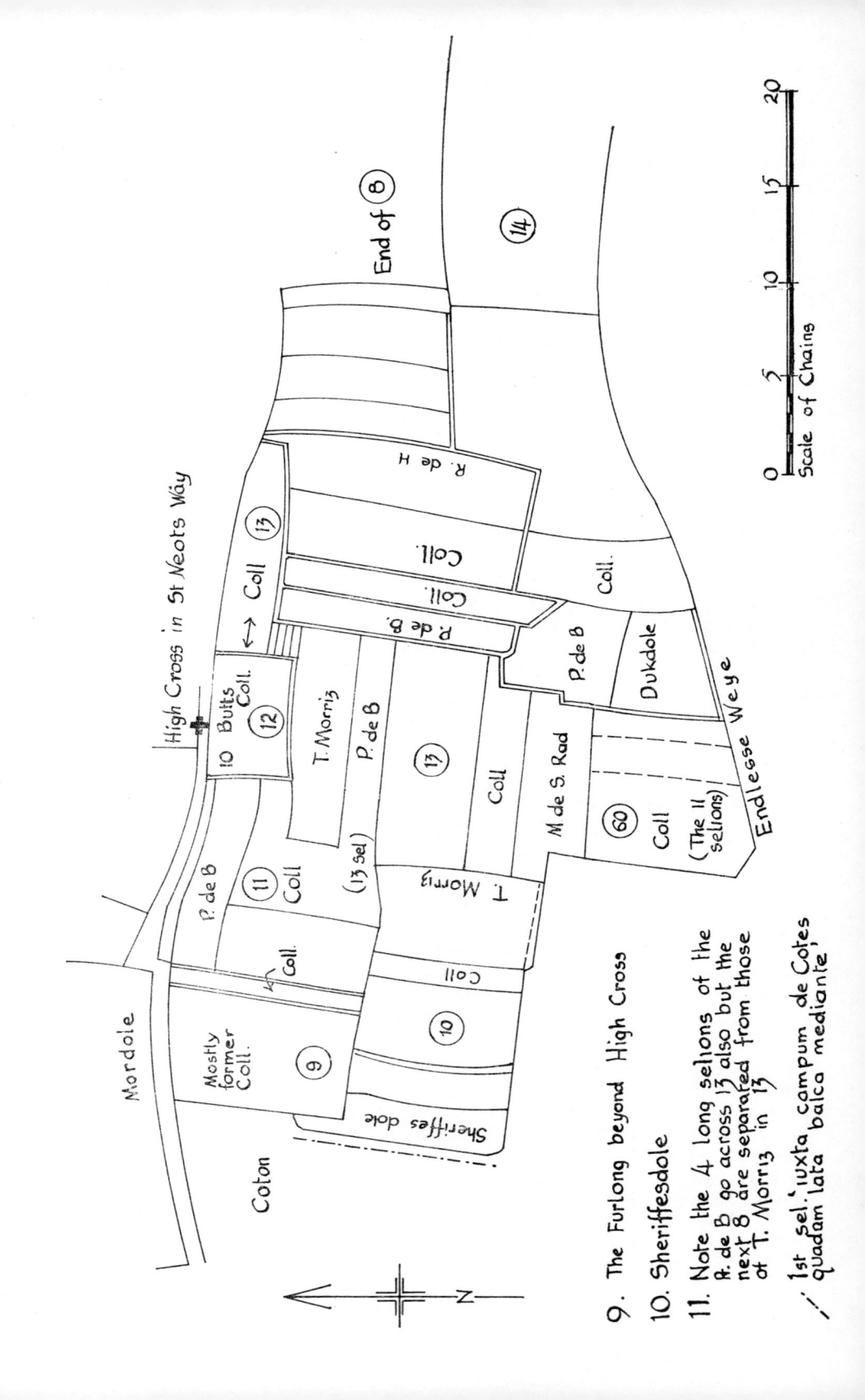

9. The Furlong beyond High Cross
10. Sheriffesdole
11. Note the 4 long selions of the P. de B go across 13 also but the next 8 are separated from those of T. Morris in 13.
   ∴ 1st sel. 'iuxta campum de Cotes quadam lata balca mediante'

the needs of the expanding community of Cambridge in corn production at a relatively late date. The topography of the fields gives us hints only as to their origin and growth, and more work needs to be done to find refutation or confirmation of current hypotheses, but a few suggestions are possible.

In the first place it seems very probable that the West Fields were shaped both by the combination of earlier systems, or parts of systems, as well as phases of assarting. Parts of Grithow may have been involved with Chesterton, the lower parts of Little and Carme with Newnham. Howes, which was certainly a separate settlement towards Girton, seems to have affected the lay-out of the field systems only by the preservation and extension of Howes Closes: it shows no sign of ever having been a centre for the development of the arable. The centre from which the further parts of Grithowfield (those numbered from 13–24) appeared to have developed, had already been much disturbed by gravel digging when the first terrier was completed. It may well have been a site of ancient settlement, for Roman remains have been found near it, on the University farm, but the extent of later workings makes it unlikely that we shall find proof, since most of the evidence must have been destroyed, or carted away and tipped into the streets of Cambridge. When the arable fields were at an early stage of development, this would have been an ideal site for herdsmen. When that part of the arable which appears in the Corpus Terrier as Grithowfield was more fully developed, an even more suitable site for the herdsmen would have been the area of the Moor Barns, just beyond the limits of the old boundary marked by the High Cross in St. Neot's way: for here the ground begins to rise slightly towards the foot of Madingley Hill, and this Moor, as it was called, cannot have differed very much from those parts of Whitwell described in Domesday as supporting many sheep and pigs, with plenty of wood for repairs. Opposite the Moor Barns, in the furlongs called "Beyond the High Cross" (Middlefield Furlongs 9 and 10) and in the adjacent parts of Coton, some rather patchy assarting would seem to have taken place earlier than in the areas of "Le Clay" (Furlongs 8, 11 and 12) or in the wet land on the further side of the Grit How (Grithow Furlong 16).

The furlong lay-out in this far Western section of the Fields suggests an unequal struggle with endemic drainage problems which made arable cultivation there impossible in the wetter climatic periods. The ploughing direction runs alternately normally and laterally in a jig-saw of small irregular patches, as if trying to catch any run-off in an indeterminate area which drains satisfactorily neither into the Washpit nor into the Bin Brook. The heavy clay soil and the readiness with which oak will grow make it likely that this was the last piece of primitive woodland to be cleared and ploughed. It was certainly the first to lapse from ploughland. In its descriptive notes the St. John's Terrier of 1566 says:

"Endless Way begins behind all the furlongs that be near St. Neot's Way next Coton Field, and at the beginning is many leys, and in winter full of water.."[1]

The version of this note quoted by Maitland is even more picturesque:

"...Beginnethe at the xj seliones of Bennet Colledge which now be lees and is called ducke pytte because it standeth full of water."[2]

1 St. John's College. xxxi, p. 29.
2 T. & B., p. 126.

Next to these eleven selions comes a block of the Prior of Barnwell which carries the marginal note in the Corpus Terrier, "Dukdole", and is thus described in the St. John's Terrier,

"14 selions of the Prior of Barnwell and they be called Duckdole, and they are all sward or grass-ground."[1]

By the sixteenth century this was enclosed with both hedge and ditch.

Another fairly extensive enclosure of chronically wet land in this area was made by Corpus some time between the sixteenth and eighteenth centuries, and is shown on the 1789 Map with a separate assessment in the schedule.

Froshiscroft or Freshcroft, at the west end of Willow's Ditch; appears in the Corpus Terrier as arable, though its name suggests that it was wet. As Hunnell's Croft, in the early sixteenth century, it would appear to have reverted to pasture.

At about the same time the nameless but important block of arable belonging to Stephen Morys, in the wet field just by Seman's Ditch, must also have reverted to pasture, thus preserving the outlines of the medieval selions for posterity. In all these cases, the key to the choice of area for reversion to leys and enclosure is drainage, combined with possession by a single owner. But drainage—or rather, lack of it—is the dominant factor; for owners were not likely to resist attempts to buy or exchange odd strips of virtually useless waterlogged arable. The persistent holder, above all the corporate owner who never dies, is therefore likely to be the foremost early encloser. Similarly, the first new enclosures on the southern boundary are in damp places, at the bend in Edwin's Brook (by Merton College) and in part of furlong 4 towards Portbridge.

One of the formative influences common to the development of many Cambridgeshire Field systems, namely, flooding—clearly operated on the other side of the fields towards the Cam. The river that we know to-day is the product of the modern canal age, but even now it can overflow its banks. It is very hard for us to imagine the medieval river, with its meanders through the marshes on both sides, and the meadows that were regularly subject to winter floods. There can be no doubt that the bluff of Castle Hill created a bottleneck for the outflow from the much wider flood plain and marshes higher up the river around Coe Fen and Sheeps Green. The area opposite the end of the Madingley Road would consequently appear to have been exceptionally vulnerable in times of extraordinary flood. Where the Madingley road now joins Queens' Road the level has quite obviously been considerably built up, as has also been the former Corpus Croft, now St. John's Wilderness. Except where special banking-up has taken place, there is no sign that cultivation was ever attempted below the 25 ft. contour. Butcher's Croft, the former Mortimer's Meadow, early enclosed, seems never to have been part of the arable. It was probably too near the 25 ft. contour for comfort. From the later terriers it would appear that up to six or eight selions at the outer margins of the furlongs nearest the river had gone out of regular cultivation by the end of the Middle Ages.

As the wetter climatic period of the central Middle Ages gave way to drier conditions, the riverside Colleges began to cover portions of the Long Green. The first to obtain a section was King's College, who, as we learn from the 1566 Terrier, bought it from the Corporation of Cambridge. It was only a matter of

1 St. John's College, xxxi. 29.

time before their example was followed by the other riverside colleges. Each college portion, as it was obtained, was surrounded by a deep ditch, necessary for drainage as well as demarcation, and the spoil of the ditch was used to make raised tree-lined walks, thus creating the familiar landscape of the famous 'Backs'. Curiously enough, King's College the pioneers in purchase, retained the ancient practice of keeping some livestock on their "backsides" when those of the other colleges had been entirely given over to walks and gardens. The townsfolk, who also enjoyed walks adjacent to the town (as we see from prospects and pictorial maps of the sixteenth to nineteenth centuries[1]). increasingly encroached upon the edges of the former for field pasture, recreation and through passage for vehicles, effectively preventing its return to arable cultivation. The choice of the 'Backs' route for the turnpike road finally limited the fields at their eastern edge to the modern line of Queen's Road.

Place and Field Names

It should be made clear at the outset that we are concerned with the place-names of the West Fields from rather a different point of view from that of editors belonging to the English Place-Name Society. Their interest has been primarily linguistic; and we freely acknowledge our debt to their work. What we have tried to do is to connect the minor field-names as exactly as possible with the topographical features to which they refer, or with the persons whose names form parts of compound field-names. On the whole, there have been few problems, since we are fortunate in having a well-written terrier and an excellent outline map. It is satisfactory rather than exciting to note that *smalmade,* the narrow meadow, appears on the maps as a thin green strip, hardly more than a widening of Sheepcote Way; or again, that the watercourse known as *Seman's Ditch* is adjacent to a block of selions "quondam Galfridi Seman"; or that *Barkersaker* in the Carmefield can be linked by charter evidence to a house in Newnham owned by the Barker family. The impression is initially of a set of names so prosaic that they seem almost inevitable. Roads, for instance, appear almost invariably as "way" compounded with the name of the place to which they lead, as *Bertonweye, Cotesweye, Eldenewenhamweye, St. Neot's Weye,* and, quite literally, *Endlesseweye,* which stops in mid-field. Pastures tend to be distinguished by shape and size, as *longe grene* and *lykylt grene.* Soil is described by its content, grit, clay, or chalk. Such names predominate and fall into place neatly on the map.

There are times when the correct derivation of a name can only be decided with reference to its topography. A case in point is *litilmer,* also written as *litilmor* in charters. We have to decide whether the place in question is more likely to be a small rough pasture or "moor" or a place with standing water, a "mere". Both may be found within a short distance of each other in the West Fields, indeed within the area covered by Grithow Field. But since *litilmer* appears to be on the slope of the Castle Hill promontory, we came down on the side of "moor". But the map cannot really help us with *Blackmore,* one of the names

1 J. W. Clark, *Old Plans of Cambridge,* 1574–1798. Part II Plans, Cambridge 1921.

for the West Fields in general, preserved only in terriers and surveys which are notoriously conservative in their terminology.[1] We can only suggest that it originally referred to the clay land of the Middle Field, which we believe to have been the last extensive area to go under the plough. There was no moor, black or otherwise, in the West Fields by the time the Corpus Terrier was written, and Madingley Moor, on the western boundary, was a moor only in name.

As the first field is dominated topographically by the gravel ridge running across it from beyond the NW boundary on the Huntingdon Road, references to the gravel, or grit, naturally appear in the field-names and form the first element of *Gyrton* (formerly *Grit-tun*) and of *Grithow,* spelt *Grythowe* in the Terrier, and *Gritho* or *Grethoe* in earlier charters. This name is given to two routes, *Grithowweye* and *Grithowpath,* leading to the ridge from the town and the Madingley Road respectively, and by the fourteenth century has been adopted as the name for the whole field. The seventeenth-century copyists spelt it *Greithowe* or *Greathowe* and tended to use the name to apply to the whole ridge; but in the earlier usage we find the term *Grithow* (or the *How* or *How Hill*) used to refer to a point at the southern end of the ridge at which the way and the path of that name met. Here was a small patch of unploughed ground called *le Greneplat,* or occasionally *Greneshed* (13c.), from which we guess that it was flat and originally roughly circular in shape, like a plate or shield (though round shields had not been seen for many generations by the time the charter which used the term "greneshed" was written).[2] It was being eroded by the activities of the cultivators, such as Thomas Morys, whose two selions, as the Terrier tells us, extended over Grithow Way and were next to a selion of the clerks of Merton which "abutted upon Grithowpath directly opposite Grythowehyll." And here, in the margin of the Terrier, (fol. 4b), is a rough sketch of what looks uncommonly like a tumulus, labelled "Grythowe". If it is an artificial mound, it would accord with Professor Bruce Dickins' comment that "the English *low* and the Scandinavian *how* generally refer to a burial mound of earth which may originally have been heaped up at any period".[3] This single "how" should not, we think, be assimilated into the general name for the settlement of the Howes, on the crest of the ridge. This name, as Reaney pointed out, is always found in the plural form, and we are not really concerned with it here, as only a very small part of that settlement (one house out of eight in a 14c. rental contemporary with the Terrier) was over the Cambridge side of the parish boundary.[4] The form of the name, *Howescroftesande,* is Girton-based nomenclature, to distinguish it from Duck End.

The name-element *how* is Scandinavian, and a Scandinavian element occurs again when we come to the other main area of higher ground, at the far SW of the Fields. Here we find the *Clint,* with its compounds *Clint Field* (in Coton), *Clint Way,* and *Clinthavedon.* Anyone used to the association of the word with

1 The same name is found in Chesterton, used for an area of rough ground behind the Castle.
2 Cf. the use in Derbyshire of 'greenplat' for the area close to a farmhouse that is not cobbled yard.
3 Bruce Dickins, The Progress of English Place-Name Studies since 1901. *Antiquity* XXXV, 1961, p. 285.
4 C.C.C. *XXXIV* 34.

a far steeper incline than anything to be found in Cambridgeshire should bear in mind that the local attitude to heights is relative. (Cambridge, like Rome, is traditionally built on seven hills, most of them only barely perceptible on a bicycle). The writer of the 1566 terrier says, "Clintway begins next Cotton Hills at a great balk that parts Cambridge Fields and Cotton fields above on a hill." The lack of a hill of suitable proportions in the immediate vicinity of Cambridge, together with the appearance of the name *Clinthaveden* in a charter of a Madingley owner[1] has led some writers, including Reaney, to equate the Clint with Madingley Hill and the Clint way with the St. Neot's way. More specific terrier and charter evidence from the Cambridge West Fields indicates that the Clint names are associated with the lesser ridge described above, which lies in part along the Cambridge Coton border. The exact position of the "Clinthaveden", or Clint headlands, is hard to determine from charter evidence, but the charter quoted by Maitland[2] which refers to selions belonging to the Hospital, "with one head on the ditch (Edwin's) and the other on the Clinthaveden" suggests that it is the rather oddly placed piece of headland separating furlongs 21 and 23 in Little Carmefield. No other noticeably Scandinavian elements occur, but the two above, taken together with *Wrangling Corner* at Moor Barns just over the Madingley boundary, give a total of three points within easy walking distance of the town of Cambridge with Scandinavian name-elements. Study of parish names alone produced nothing nearer than Toft.

The only other part of the Fields above the 50ft. contour level is the central part of Middle Field called *Aldermanhill*. Charter evidence pushes the use of the name further back when it could reasonably be supposed to relate to a town or gild official, and we suggest that it was originally "ealdorman" hill, and a designation of the late Saxon period. But whether the hill was important for defence, for reviewing of the fyrd or the scene of forgotten military activity, for instance, or in any way connected with Edwin of the ditch that bears his name, can only be a matter for conjecture. *Edwin's ditch* ran through the small valley between *Aldermanhill* and the *Clint,* which in the early Merton and St. John's charters is simply called *Dale* or *Le Dale,* while the way along the Coton boundary leading to it is called *Le Daleweie* or, less commonly, *the Daleride*. But by the time of the Corpus Terrier it is *Dedale*, and in charters also *Deddale* and *Dedmannisdale*. Whether any foul deed had actually been committed there we cannot say, any more than we know about the reason for the similarly named and placed *Dedmansway* between Barton and Grantchester. Probably folk tradition held that low-lying and consequently shadowy ways were dangerously liable to be the scenes of crimes and of hauntings. When eighteenth-century rationalism had banished such superstitions, the name was misleadingly written as *Dedole* and entered on the 1789 map in the wrong place.

As starkly unimaginative as *Le Dale* is the early nomenclature of the chief watercourse in the West Fields. Both before and after crossing the Cambridge boundary it appears to have been known as *the Brook*. As such we find it in charters of the early 13th century which refer to arable land "super le

1 C.C.C. *XVII* 2.
2 *T. & B.,* pp. 172–3.

Broc", and in Coton parish it long remained as simply the Brook. But the early name for the field within the curve of the brook on the Newnham side was *Binbroc,* "within the brook", and in the St. John's charter quoted by Maitland[1] the arable here is referred to as *Binnebrok,* while another more westerly piece which abuts on Edwin's Ditch is *Butebrok.*

Maitland comments, "observe the contrast between Binnebrok and Butebrok". We should point out more explicitly that "binn", within (as in German 'binnen') is contrasted with "bute", in the sense of without or beyond (as in the cognate Dutch "buiten" or the Yorkshire 'baht'), so that *Butebrok* is what is translated as the Latin charters as "super le broc" and is the oldest name for the arable on the left bank of the Brook.[2]

The use of *Binbroc* for the whole of the arable on the right bank of the stream was current by the mid-thirteenth century, and by a back-formation the name was applied to the stream itself, which had become *Bynnebrookes* by the time of the Corpus Terrier. As *Binn Brook* it has remained. The field, as we have seen, shortly thereafter received a new name *Carmefeld* derived from the Carmelites and their *Carmedole,* so that the origin of the name for the stream in the field was forgotten. Of the other water courses, only *Bradderussh,* a tributary of the Girton Washpit Brook on the western boundary of the West Fields, has a descriptive name, reminiscent of the picturesque names found near Oxford. The parallel watercourse above it, probably only seasonal, appears to be nameless, though the green strip through which it flows is called *Kynchismade.* We can throw no light on the first name-element. The remaining watercourses are called "ditch" as often as they are called "brook" (e.g. *Willowesdich*), and we are reminded that very few of the streams crossing the arable areas of southwest Cambridgeshire have been left untouched in their original courses by the time that we have written evidence about them. The Newnham mill-cut, the Cam itself, and probably also the lower end of the Binn Brook have been harnessed to the working of mills. The smaller watercourses have been straightened, deepened or turned round the edge of arable furlongs to drain or to form boundaries to them. It is therefore not unusual to find the names of these semi-artificial watercourses compounded with a personal element.

The names for the two main tributaries of the *Binn Brook, Seman's Ditch* and *Edwin's Ditch,* have already been mentioned in connection with the persons to whom they refer, the former well documented, the latter unknown. The older form of Edwin's Ditch is *Edinebroc,* and in a Merton charter *Ediwei* brook.

Another completely unknown person whose name attaches to a notable man-made feature of the West Fields is "Custe" of *Custesbalk.* We have not been able to discover any earlier forms of this name than that in the Corpus Terrier. Most of the other names with a personal element are given to arable strips and associated with former owners. *Godivesdole, Herwardaker,* and *Godwinesrode,* all suggest pre-Conquest family names. Other names are connected with persons living or only lately dead at the time of the Corpus Terrier. Such are *Porthorsdole, Thorpiscroft, Shermannisrod,* and *Barkersaker.* After these,

1 *T. & B.,* p. 172.
2 Professor Dickins points out that the same contrast is retained in the Scots names for outer and inner rooms, known as the *but* and the *ben.*

in fifteenth and sixteenth century marginal notes, we have *Spaldyng's Close, Sheriffe's dole, Hunnell's Croft,* and *Colys Cross.* A further group of names is connected with official or corporate owners, usually of a nearly date. Such are *Erlesdole* (the Earl being the Lacy Earl of Lincoln; see p. 73) and possibly *Aldermanhill,* while ownership by a town gild is suggested by *Tunmannisaker* and the two *Gildenakers.* The connection of the Carmelites with the house at Newnham and the consequent naming of the *Carmedole,* and subsequently the *Carmefield,* are well authenticated. The Franciscans have left their mark in *Le Cundyt* on Howe Hill. Another religious house which has very probably left a name in the West Fields is Barnwell Priory; for the building to which account was rendered was, and is, known as the Chequer, and the name *Le Chekker* is given to an area in the far west of Grithow Field where Barnwell ownership was paramount. *Gilisrod* was owned by St. Giles' Church.

An interesting connection can be shown between *Le Lampeaker* in Old Newnham Field (Q1) and the land which the Hundred Rolls tells us was given to the Hospital by Hervey fitz-Eustace to provide a lamp before the sick at night.[1] The Hundred Rolls also helps us to connect *Karlokaker* with a certain Richard Karlok recorded in it, rather than directly with the noisome arable weed charlock. But who can say whether or not the name "charlock" was first given to Richard or one of his family in derogatory reference to the condition of their arable holding?

By the time of the Corpus Terrier the arable seems to have been used for corn to the exclusion of other crops, but the thirteenth century *Linrode* in Dedale suggests that flax was once grown here. Unlike Grantchester and some other neighbouring parishes, the West Fields have no *Benelond* or similar name. By EldeNewnhamweye is a *Dufowshyl,* or dovehouse-hill.

The Claypits and Gravelpits need no further explanation, though it may be noted that the results of both these traditional diggings can still be seen, the former at the University Farm, and the latter at the site of a controversial proposed hotel on Castle Hill. Close to this site another less reputable human activity has left us the name *Hor Hill* or *Hores Hill,* variously mistranscribed by later writers as *Horse Hill* and *Hare Hill.* A similar name, to the South of the lower town, provides yet one more instance of the dual functioning of the town of Cambridge. Both were traditionally disreputable areas until well into the present century.

There is the usual crop of names referring to the nature and appearance of the soil, *Le Cley, Blakaker, "Spony* (spongy?) *aker", Peper* (pebble) *dole,* and *Stonigrund.* Wet places are connected with frogs or ducks, so that we have *Dukmere* and *Dukpytte,* and also *Frosshescroft* and the *Frosshelak.* The former, near Vicar's Farm on the Madingley Road, was corrupted to *Freshcroft* when the older form "frosh" was replaced by "frog". The latter was even more perversely corrupted to *Frostlake,* a name retained by one of the cottages in Malting Lane, though the lake itself has been incorporated into the gardens of Newnham College. An interesting name on Castle Hill, near the Chalkwell and probably affected by a run-off from that springline, is *Muscroft.* The first element of the name would appear to be that found in the Lancashire "moss" or the Swiss

[1] *Rotuli Hundredorum,* ii, p. 361.

"muesli" (a little swamp). Although this croft is given as arable in the Corpus Terrier, it was soon afterwards enclosed and withdrawn from the general field system.

Finally there are the numerous names referring to the shape or size of the arable strips, as *goredaker, cutted rod,* or to the shape and position of the piece of field, such as "*the Hearne* called *Le Erbeer*", in a corner formed by Willows Ditch. Such a descriptive title as "the harbour" is unusual. As has already been stressed, by far the greatest number of the minor field names are entirely prosaic and relate, predictably, to arable cultivation.

Nomenclature of the Great Fields

The nomenclature of the fields as given in the Corpus Terrier is the one normally used in this commentary, and was indeed consistently in use in other terriers from the fourteenth to the nineteenth century: that is to say, Grithowfield (sometimes corrupted to Great How), Middle Field, Little Field, and Carme Field. The last had acquired the alternative name College Field by the sixteenth century,[1] and was so called in the St. Giles Enclosure Award. The Corpus Map of 1789 used a system of its own, putting Little Field into the section of Middle Field (furlongs 8b–14) that was ploughed with Carmefield, and Dedole into the old Little Field. A new "Little Carme Field" appeared in furlongs 21–23 of Middlefield. This seems to be aberrant: the Jesus College Terrars[2] from the sixteenth to the nineteenth century maintain only the four fields identical with those of the Corpus Terrier.

It is clear that these names, though well-worn by 1800, had not been thus applied from time immemorial. The Carme Field, for instance, which Maitland[3] correctly attributes to the Carmelite brothers, whose house was based in Newnham for forty years up to 1292, must have had another name before the mid-thirteenth century. The Hundred Rolls of 1279, besides the Carme Field, mentions only the Cambridge Field and the Port Field. Maitland, with characteristic acuteness, comments, "I am not sure that the whole of the Western Fields were not known as the Port Field. I have seen a charter in the Archives of Merton College[4] which seemed to point in this direction. To find a Portfield or Portmeadow outside a borough is not uncommon.[5] The little bridge where the Bin Brook enters the Cambridge Fields was apparently known as Portbridge." He might have added that the road coming over this bridge which becomes Old Newnham Way is called Portway in Grantchester Parish. Maitland was partly right in his supposition, in that the term Portfield seems to have been used for a very large central portion of the West Fields, but where it has been possible to check the provenance of identifiable parcels of land back to the charters of the early or mid-thirteenth century we cannot find it used in either of the

1 St. John's College, xxxi, 24.
2 Jesus College EST. 4.2., EST. 4.3. & EST 4.7.
3 *T. & B.,* pp. 121, 122.
4 Probably Merton College M.1580.
5 A comment possibly made with Oxford in mind.

oldest areas of cultivation, based on Castle Hill and Newnham respectively, in the Grithow and Carme Fields.

In the central area the terms "Portfield" and "Cambridge Field" seem to be more or less interchangeable, but while "Cambridge Field" is very definitely used for all the northern sector of the fields, and indeed is the normal way of indicating property in what is later the Grithow Field, the term Portfield does not seem to have been used in furlongs 1–15 of Grithow at any time. The most northerly use of the term "Portfield" we have yet discovered in charter evidence is the area of the St. Neot's Road in the far west, beyond High Cross. Here the description "at Morhyl" would seem to refer to the area on the north side of the road towards Moor Barns or Madingley Moor, where a property of the Clerks of Merton is found in furlong 19.[1] In the southerly direction all the furlongs of Alderman Hill, including those abutting on to Edwin's Ditch, were reckoned in Portfield, but we have not found it used for the area south of Edwin's Ditch or in the Carme Field.[2]

In the older charters the names Grithow or Gritho, Middlefield and Little Field do indeed occur, but they refer to a much smaller area than that covered by the fields of the same names in the Corpus Terrier. The earliest charter reference to Grithofeld in the Corpus Charters is dated 18/19 Henry III (1234), but the more usual form of reference is "in the Cambridge Field towards" Gritho, Greenplat, Litlemer or Howescroftesende.[3] Similarly Smalemadwe, Coteweiesende, Aldermanhylle and Edwinebroc are used for reference in the central area.[4]

The southwestern parts of Middlefield, called Le Dale or Deddale, had a separate existence from early times, and come under a sort of sub-heading even in the Corpus Terrier.

The earliest occurence of the name that we have found is in an early charter of Gonville and Caius College, and refers to eight selions in "Le Cley" in "Middlefield" (circa 1240).[5] This is one of several indications that "middle" was at first applied only to this sticky clay patch, and we think it possible that it was originally thought of as intermediate between the older cultivation of furlongs 1–8a, and the assarts at High Cross (furlongs 9–12), rather than as intermediate between a northern and a southern set of fields.

That the northern and the southern sections of the West Field were originally separate is clearly indicated by their topography and lay-out, as well as by analysis of ancient ownership and tithe system. The evidence of nomenclature in no way contradicts this supposition, and we would go further and say that it suggests that the ancient limit of the southern system was Edwin's Brook to its junction with Bin Brook. The arable did not anciently extend to the furthest limits of this area however, for Maitland's St. John's Charter refers to "le longum pasture iuxta Edinbroc".

1 Merton College, M.1580
2 C.C.C. *VII* 20.
3 As in Merton College M.1606. C.C.C. *VII* 13–18 and St. John's College, quoted *T. & B.*, p. 172.
4 C.C.C. *VII* 7, 8, 20 & 22.
5 G. & C.C. *XIII* 1a.

The earliest references to Littlefield which we have found, in the Cartulary of Sir John de Cambridge and in a Corpus Charter dated "m cc lxxx primo", are to a piece of land in the third furlong of that field, and we have yet to find any early document in which "Lytelfeld" is used for parts of Littlefield north of Edwin Brook. The most usual way of describing the southern part of Littlefield in thirteenth century charters is "super le broc", and similarly the fields of the southern system are simply described as "in Binbroc", or, more rarely, "in Binnebroc feld". The Carmedole itself, which later gave its name to the whole southern field, was in the first half of the thirteenth century predictably called "Spiteldole".[1]

Extensions of the southern fields into the south and southwest were described as "apud Purtebrigge", (e.g. furlong 4); and the mysterious Clynthavedene, quoted by Maitland, and also found in some charters, would seem to refer to the far south western corner, later known as Little Carme Field, adjacent to the Clynt Way.

The southern fields as a whole were frequently referred to as the Newnham Fields, as in the Cartulary of Sir John de Cambridge, "in campis de Newnham"[2] and the portions closest to the hamlet (furlongs 1 and 2) were sometimes called "Eldenewenham" field. Surprisingly, the term "Newenham Croftes" applied to the area around furlongs 2, 5 and 6, is found as early as the mid-thirteenth century, long before the furlong was withdrawn from the arable.[3] The name must therefore be older than the earliest enclosure here, and may well date from a very early stage in the development of the fields. Newnham was one of several small hamlets in the vicinity of Cambridge, consisting of cottages with crofts, which had not achieved parochial or manorial status by Domesday, but which were awkwardly split by the boundaries of larger neighbouring units (cf. Howes and Cotes). One cottage with its tiny croft, right out in the fields in a manner reminiscent of Thomas Atchurch of Howes, is found in the hands of the Pestour family for several generations. The original deed of grant comes from the lords of Grantchester, but by the fourteenth century the cottage is reckoned as being "at the end of Newnham village" and pays dues to the Mortimer manor, although it is on the "wrong", or Grantchester, side of the main road. Grantchester records indicate the presence of other crofts and orchards reckoned fully in Grantchester parish, and it would seem that all these crofts predate the expansion of the arable from both Cambridge and Grantchester, which had nearly but not quite engulfed them by the fourteenth century.

But the main hamlet of Old Newnham clustered near the mill was more important than the crofts, and it is here that we find the houses of the owners of strips in the "Fields of Newnham". The term "Fields of Newnham and Cambridge" persists in the charters to a surprisingly late date. In the series of Mortimer terriers[4] which cover property on both sides of the river, the normal term is, "Fields of Cambridge, Barnwell and Newnham". Interestingly, the

1   G. & C.C. *XIII* 23 and *XIII* 18; C.C.C. *VII* 1, 3, 4.
2   C.C.C. Sir J. de C. fol. 33.
3   C.C.C. *XV*. 12.
4   G. & C.C. *XIII*. 23.

sequence followed by all the Mortimer terriers takes the form of a circular tour round the town, in the earliest terrier, starting in Newnham, crossing the river to the Trumpington Road area, thence to Barnwell Fields, and finally round to "Cambridge Fields" based on Castle Hill. Such a sequence adds weight to the argument that there was originally a clean break between the Newnham-based arable and the Castle Hill-based arable.

Terrier writers are notoriously conservative in matters of lay-out and sequence, and the Mortimer terriers are parents of a family of terriers, quite different from the Corpus "tithebook" type, and easily recognisable from their heading, "Fields of Cambridge, Barnwell and Newnham", and from their adherence to a circular sequence that ends awkwardly in the middle of Middlefield. Later terriers, however, reverse the anti-clockwise circle starting at Newnham, to a clockwise circle starting on Castle Hill. To this family probably belong the seventeenth century terriers of "University Lands" preserved in the University Archives. The essentials of a terrier are also embedded in the charter of Richard Dunning from the Mortimer deeds.[1] This charter is an excellent example of the way in which a mixture of old and new styles of nomenclature are combined for identifying various parts of the fields at the beginning of the fourteenth century.

This brief survey of the nomenclature applied to the Cambridge West Fields should be sufficient to indicate that although field names may be very ancient, the area of ground to which they are applied at different dates can expand, contract, or shift, or the name may be replaced altogether. It is therefore just as important for names found in eighteenth century maps and terriers to be checked by older documents for changes in their exact application, as for changes in their form. Both are likely to have been subject to popular rationalisation, or plain misunderstanding on the part of clerks, where the words they are copying have through the passage of time or their ignorance of the area ceased to be meaningful to them.

1   G. & C.C. *XIII* 1b. See Appendix B.

HINTS OF ORIGINS

Early Forms

When we put together the clues supplied by topography, place-name evidence and ownership, can we detect anywhere in the West Fields the signs of more ancient patterns of arable cultivation which have been incorporated into, or influenced, the medieval lay-out? One indication would be the existence of small squarish fields, sometimes miscalled "native" or "Celtic" fields, or more accurately, "blockfluren", of the type generally associated with very early agricultural patterns. The known archaeological evidence from the Upper Town and its environs suggests considerable activity in the Roman and immediate post-Roman periods, followed by decline and stagnation until Saxon times. It would therefore be reasonable to suppose that any arable cultivation which persisted in the "hinterland" of the Upper Town, in however rudimentary and intermittent a form, would tend to be on a small scale and probably within the limits and conforming to the existing boundaries of former ancient fields. We would suggest three places where we may possibly have traces of such residual older fields.

The first and most obvious place to look is in the immediate vicinity of the Upper Town, while bearing in mind that only half its hinterland lay in the West Fields, and in some ways the less easily worked half, for the bluff now topped by the Castle Mound falls away steeply into what was "The Marsh". Moreover the area has been disturbed by claydigging at Ashwykston, gravel digging on Pound Green, and by various phases of defensive works, the latest of which was in the Civil War. In what remains there is a noticeable squared pattern in Muscroft, the croft of St. John's Barns, the piece of 'Hospital' cultivation at the top of Q.3 encroached upon by the claypits, and perhaps at the top end of Little Field, in Qs.8 and 9.

The next point to consider is at Howes Crofts, where an ancient settlement straddles the parish boundaries. The greater part of this settlement lay in Girton parish and is therefore outside our terms of reference, but a small part just comes into the far western fringes of Grithow Field. Here the last selions of Q.4 are of a different length from the rest of the furlong, and are held by the owners of the houses in Howes, Thomas Jekke, Robert Rye and the like. Their holdings are grouped without intervening balks into squarish units. More striking is the self-contained block of Q.14, all held by owners of Howes' tenements. These parts of the West Field were withdrawn from the general arable and enclosed by hedges soon after the writing of the Corpus Terrier. The hedges, which follow the outlines of the former blocks, continued to form a markedly squared pattern on later maps, even more impressive when the Girton side is considered together with the Cambridge parts.

The third area, that of the Newnham Crofts, is the most striking of the three here considered. The whole of this part of the Carmefield (Qs.1–6) is based on a grid lay-out, which extends across the boundary into Grantchester parish. Even on the latest maps the outlines of the furlongs are hardly affected by the S-bends characteristic of the Middle Field. This may be in part due to the restraining influence of the two great ancient balks, Long Balk and Custes Balk, but there is also a furlong boundary, at the edge of Q.2, which is just as straight. The persistence of the grid lay-out, even though cut diagonally by Old Newnham Way in Q.2., makes one think that here, if anywhere in the West Fields, we may have the remains of a pattern modelled on centuriation.

The St. John's 1617 version of the Terrier gives the following note about the cropping practices and extent of the crofts area of the Carmefield, "There are in this field certain crofts which be every year land and never left fallow but at the will of the owner. These crofts are contained in the 1, 2, and part of the 5th furlonds and all the 6th furlond croftland and no more". The area so designated coincided almost exactly with the area of grid lay-out just mentioned. It has been argued that the cropping practice peculiar to the crofts was the beginning of a break-down and withdrawal from the general seasonal shift, leading ultimately to enclosure. This does not make sense in view of the fact that the crofts were not separated from the rest of the Carmefield by hedge and ditch or by any visible barrier, either at the time of the Corpus Terrier, or in 1617. Indeed, by the time of Parliamentary enclosure, only a small portion of Q.5 had been thus withdrawn. It makes more sense to see the practice of the crofts as the survival from an earlier period before the three-seasonal shift for the whole West Fields had been established. It ties up with the very ancient use of the term "Newnham Crofts", found in the oldest charters we possess, some of which preserve the individual names of crofts (such as the thirteenth century "Asshemannes Croft" (CCC XIV.1). Moreover, analysis of ownership points to the survival of some groups of selions held by one owner who appears to have no other holdings (e.g. Geoffrey Wardeboys in Q.2.) in the fields and to a link between such holdings and a tenement in Newnham hamlet. In this respect the pattern has similarities with the Howes/Girton ownership.

It is to be noticed that at Newnham Crofts, as at Howes Crofts, the older cultivation sits astride the later parish boundaries. The inference is that these smaller patches of settlement and cultivation were deliberately partitioned between those neighbouring larger ones which had parochial status when the parish boundaries came to be drawn. We would suggest that in a region of numerous settlements of varying sizes, such as this part of Cambridge, the presence of a "stepped" parish boundary is as likely to be indicative of a patch of older cultivation belonging to a small sub-parochial settlement as for any of the other reasons generally given. Such small settlements could be either the remains of very ancient ones, or the embryos of newer ones, and in studying the vexed question of the rise of "The Cotes" to the later Coton parish, the position of the "stepped" boundary of the West Fields and Coton by the fourteenth century should be taken into consideration. The history of the development of the West Fields shows clearly that the arable was not uniformly extended from age to age like ripples from a stone thrown into pond. It grew from various centres at different rates for a variety of reasons. These growing-points would usually consist of the main one, which gave its name to the parish or area, supplemented by the subsidiary ones, often on the very perimeter of the parish area. Infilling was the final stage and was not completed until relatively late in the Middle Ages, perhaps in some cases as late as the thirteenth century.

The Great Plough-Up

One of the attractive features of the Corpus map of 1789 is the way in which the selions belonging to the College, with their attendant balks, are faithfully drawn straight or curved, and that this is not just the whim of the cartographer is borne out by comparison with the furlong outlines of the draft Enclosure Map. The curved selions of the reversed-S shape, associated with the heavy eight-ox ploughteam, are most marked in the central area on either side of the Madingley Road. We find them shown on the map, for instance, in Q.5 of Grithowfield, while in Q.15 the actual selions are preserved in pasture for us to walk and measure.

In Q.15 of Middlefield they are again clearly shown on the map. Can this tell us anything of the relative times at which this area went under the plough? We can at least say that the very wide furlongs in which these curved selions are found can only have been set out by a community fairly advanced technically and strong in numbers of plough-teams. The soil is in parts intractable: the numbers of selions to a furlong runs almost into three figures. This all points to a relatively late stage in the development of the West Fields. It has been suggested that it is the Scandinavians, with their tendency to rate everything in terms of arable, the "ploughland", that we owe the origins of the policy of ploughing up in the parishes of southwest Cambs. from east to west and from north to south—producing the enormous cornfields that so surprised Maitland. Indeed, there can have been little left to plough but inconsiderable trifles of marginal land, strips along the stream margins here, or encroachments on the pasture there, (such as Robert Long's strips in Braderushe, and odds and ends usually found belonging to the Prior of Barnwell in the Western areas), after the great plough-up of the Central Area had taken place.

It is, of course, equally possible that while the first large-scale attempts to bring the central area under the plough were initiated and perhaps made necessary by the influx of Scandinavians, (see the Scandinavian names of the Clint and the Grithow), the final assarts were not made until the eleventh century, or even later. The increased need for supplies, which must have accompanied the presence of a military garrison in the tenth and eleventh centuries, was followed by the economic incentive of a flourishing Cambridge corn export market in the twelfth century. In the thirteenth century came a rapid development of other trades and activities, bringing an increasing demand for bread and beer from the town, the religious orders and the new scholars. So the impetus was to plough and plough until at least the turn of the thirteenth to fourteenth centuries. It was an impetus which once set in motion was hard to reverse, until Nature revolted in reduced yields. Even then stagnation with minimal retreat was the result, rather than radical rethinking.

The policy of the great plough-up produced the twin problems of hay and manure. By the time the Corpus Terrier and its accompanying documents were written, it is clear that a fairly sophisticated economy prevailed in which proceeds from the sale of corn were used to buy hay from the Fens for winter use, and there are records of considerable agistment of stock on summer Fens pastures, or on the commons and rough pasture of other parishes. Cattle appear to have been moved a considerable distance over green roads or by water. By these means the problem of hay was dealt with, though it seems certain that agistment left the arable badly under-manured. The solution to this adopted by the local farmers of the sixteenth to eighteenth centuries was the purchase of kitchen waste and night soil from the town, particularly from the Colleges. To this practice we owe the legacy of numerous small sherds in the soil, familiar to gardeners in the Cambridge suburbs and visitors to the Botanic Garden.

But the men who first ploughed-up so extensively had not this range of economy and we are left wondering how they expected to keep all those plough-oxen and the cows which must produce their replacements through the winter, when they had left almost no meadow and very little pasture outside the stubble. One suggestion is that they kept the numbers of wintered cattle very low, a solution which would be particularly bad for the fertility of the soil. Another suggestion, put forward by Professor Postan, is that the newly laid-out arable, with its "wasteful" balks and headlands, was intended from the first to bear its own hay, in the balks, and, at an early date, in the headlands also. This surrounding of strip-patches of corn with hay is (or was until very recently) still practiced in agriculturally backward parts of Swabia. The corn is normally wintersown and the hoeing, etc, carried out before the rapid growth of hay in spring and early summer. Cutting the hay gives access to the corn. It seems reasonable to suppose that this could have been the ancient practice here, and only with the increased pressure to produce more corn and the possibility of buying hay were the headlands and odd corners ploughed, and the balks, as it appears from the Terrier, encroached upon by odd selions here and there.

Cambridge as a Shire-Town—The Shift to the Port

The erection of Cambridge into the head and name-town of a shire of fourteen hundreds had taken place by the end of the tenth century, though the exact stages in the process are not entirely clear. What seems indisputable is that it was part of a policy of selecting certain towns of military and strategic importance to become focal points of local administration. Anciently the Romans, and more recently the Danes, had selected Cambridge in the same way. (A.S.Chronicle for 875, Danes wintered in Grantebrygge.) It is likely the Cambridge subdivision into hundreds followed initially along the lines of the Danish wapentakes. But if previous conquerors had looked primarily at the military potential, they had not entirely neglected the trading potential, and likewise under the later Anglo-Saxon monarchy the shire towns had to be "ports" as well as "burghs". The Danes had developed the riverside trading area of Cambridge across the bridge and the main market area lay again beyond this in the centre of the lower town. To meet all the requirements of the new "burgh and port" Cambridge had to unite her dual (if not triple) component parts—a sort of Buda Pesth of mid-Anglia.

This union was followed by an administrative adjustment which resulted in the effective dismemberment of the area formerly forming part of the environs and certainly of the field system of the old Upper Town. The settlement of Chesterton, which from its name alone can be seen to have had a close associa- with Bede's "Grantacaestir", and whose fields formed an integral and possibly the more important part of the area farmed by the owners living in the upper town, was henceforward erected into a place in its own right. Not just into a parish, but into the name-parish of a hundred, a curious disjunct hundred which bears every sign of having been a contrived and artificial creation. No doubt the object of the contrivance was to strengthen royal control, for from the surrender of the Danish army to Edward the Elder in 921 Chesterton had been a royal vill. The Normans, in their brutal way, accentuated the division when they demolished a large part of the upper town for their castle, and set it administratively within the bounds of Chesterton.

So while in one way the new arrangements brought importance to Castle Hill as the seat of a shire, in another it brought dismemberment and partial destruc- tion to the old upper town. Moreover, the growing economic importance of the area round the market of the lower town was to shift the balance of power and wealth in favour of the latter. After Harvey Dunning there were no great men in medieval Cambridge who drew their wealth and prestige from their lands in the fields of the upper town.

The shift in balance was reflected in the ownership of the West Fields. The more recently developed central parts of the West Fields became, not an extension of the preserves of the Castle Hill farmers, but the land of the joint community, the Port Field, with the majority of owners living in the lower town. By the end of the Middle Ages, only collegiate owners, (John's, Corpus, Clare and of course, Merton) who had become the successors of earlier substantial owners and so fossilised their holdings, retained farms of any size based on Castle Hill. Those other substantial owners who were townsmen set up their town houses or "manor" houses in the lower town—the Harleston manor in Thompson's Lane, Cotton Hall, and the Cayley manor, for instance. And though the Long Green was never called Port Meadow, or Port Green, as well it might have been, it was the Town Burgesses, sitting in the Guildhall, who finally decided that it was in them that lay the vested right to sell off portions to Colleges in the sixteenth century.

LORDSHIP AND OWNERSHIP
Lordship in the West Fields
A. Merton

The question whether there was ever a lordship of the West Fields in the manorial sense, or whether there had ever been a "Lord of Cambridge-on-the-Hill" is for us, as for Maitland, a matter of purely academic interest.[1] But Maitland reminds us that at the time of the proposed enclosure of the West Fields, in 1802–3, the matter was one of sufficient contemporary importance to require settlement by a court judgement. The verdict itself, given in the Cambridge Guildhall by a jury of merchants from London, has very little historic value, since in the circumstances it was bound to go in favour of the town Corporation. More important are the views of the unsuccessful contestors, Merton College, Oxford, and St. John's College, Cambridge. Maitland quotes a certain, "Mr. Marsh, who was managing the Bill on behalf of St. John's College" as reported to have said "that he had always understood the manor as belonging to Merton College, Oxford; that at least the only court baron held in the parish of St. Giles had been holden by the said College from time immemorial". And in the court proceedings of January 1803, Merton claimed lordship of the waste, pointing to the faldage their tenant had over the whole area when corn was off, and to the right of going on lands of those who did not hold of the college: St. John's had a right equally extensive.

The claim of Merton College to hold a court in its Cambridge Manor and to be exempt from suit in other local courts was one which appears to have been contested soon after it was made and from time to time throughout the Middle Ages, and was only finally made good by a royal grant in the Civil War.[2] This claim to exemption was based on a grant from Henry III, made in 1271, giving the Scholars immunity from suits of the county, hundred, wapentake etc. in respect of their scattered possessions.[3] It was put into effect almost immediately in respect of property in Chesterton, and probably also in Grantchester. To exercise immunity of this sort with regard to peripheral possessions does however seem to suggest the existence of a manorial or quasi-manorial court based on the nucleus of the Cambridge property. The court was always held in Merton Hall and was presumably descended from such rights as had been possessed by the Dunnings, the previous owners of Merton Hall and of the main Cambridge properties of Merton College.

It is to the Dunning records, preserved in Merton College, that we must therefore look for the antecedents of Merton lordship and much of Merton

1   *T. & B.*, pp. 201–2.   See also *Narrative of the Proceedings on the St Giles's Inclosure Bill.* London 1802.
2   J. M. Gray, *The School of Pythagoras,* C.A.S. 1932 pp. 23–4.
3   *Rotuli Hundredorum,* ii, p. 407.

ownership in the immediate Cambridge neighbourhood, and in particular in the West Fields.

Maitland was quick to recognise that "the estate which had about it the strongest manorial character was the Merton estate", but that this estate was not exclusively or even primarily in the fields of Cambridge and that the large Grantchester section did not come from the Dunnings.[1] To prove the continuous existence of a manor court for the Merton property from the date at which the College entered into possession does not, therefore, of itself, prove anything certainly about the position in West Fields before that date, though it may act as a pointer to where the clues are to be found.

Maitland considered the Dunnings as "greatest of all" the families holding property in the Cambridge fields. He briefly outlined their descent from the "man who had no name but Dunning" and sketched the rise and fall of his descendants. He felt that a more detailed study should be left to an alumnus of Merton College, whose records he had not had a chance of examining closely. He could not quite make up his mind about their status; "at first we might be inclined to claim for the Dunning family some exceptional position in, or even some sort of lordship over the town", but decided that "we can only give them a primacy among equals.[2] In drawing this conclusion Maitland was not, we think, making sufficient distinction between the position of the Dunnings as townsmen and officers of the borough, and Dunnings as manorial or quasi-manorial lords in respect of their landed holdings in the vicinity of Cambridge.

As civic dignitaries, the Dunnings, while they supplied Cambridge with the first mayor to make written use of the title, were not among the wealthiest of the burgesses. In the early thirteenth-century tallage lists, published by Maitland, their contributions are exceeded by several others, tanners, smiths and corn-traders among them; they do not appear at all in the amercements of the traders of 1177. Houses and rents they may happen to possess in the town seem to be incidental and to be appendages of certain bits of land. Down in the borough, Hervey fitz-Eustace had, as Maitland said, "a primacy among equals"—or even less than that. But up on Castle Hill, as local landowners, the Dunnings had an exceptional position, and possibly some sort of lordship.

The activities of Hervey fitz-Eustace Dunning, as he contested his rights in the Courts at Westminster, are recorded in the Feet of Fines and Curia Regis Rolls from the reigns of Richard I and John.[3] The records of Merton begin with a concord of 11/12 Henry III (1226) and thereafter fill out the story in detail until the completion of the sale to Merton College at the end of the century.[4] Even from the early scattered references in the central records, it can be seen that Hervey's opponents are drawn in roughly equal proportions from the wealthy burgensic families of Cambridge and from the local landowning upper classes. The area in which he contested his rights was partly in the town, for tenements only, but more often for landed interests based fairly equally in the Fields of Chesterton and in the West Fields of Cambridge. One has the

1 *T. & B.*, p. 74.
2 *T. & B.*, pp. 164–6.
3 J. Hunter, *Fines Sive Pedes Finium*, Record Commission, 1835, pp. 260–84.
4 Merton College muniments, M.1555 in particular.

impression, which increases as it is backed by more detailed study of the later documents, of a family curiously and uneasily poised, socially between the town burgesses and the descendants of the feudal aristocracy, topographically astride the Huntingdon Road, lords neither in Cambridge nor in Chesterton.

The early concord to which reference has been made may act as an illustration. It was made in full Halimote of Chesterton before Lawrence, Prior of Barnwell, as lord of the manor of Chesterton, between Hervey fitz-Eustace and William, son of Hugh de Chesterton. It appears to be a tying-up of loose ends left from an earlier dispute of 2 John touching the division of the inheritance of Dunning and the rights of Sir Robert de Chesterton in certain property in Chesterton. Each party quitclaims to the other the tenements he holds "de feudo Dunning domini et de feudo Roberti militis". It goes on to arrange the rights of faldage, to be equally divided month and month about, and the rights of fishery "cum sanone", and the rights of free bull and boar, each party to have a moiety. The important point about this concord and about the dispute it was to settle is that, while the Prior of Barnwell has become the Lord of the Manor by a charitable donation, he has within his jurisdiction two ancient and substantial fees, each with its own considerable rights, overlapping and not clearly demarcated from the other. Of these two fees, one is that of the family of Sir Robert de Chesterton, the feudal, military tenant of Chesterton; the other is the fee of "Dunning the lord", a family said to have owned property in the area by ancient descent since before the Conquest. Are such rights as they possess of equally ancient descent?

The fact that Dunning is reckoned as a lord only and not as a knight is interesting in another context. The most flamboyant of the Dunnings, Hervey fitz-Eustace, is recorded as having attempted to press his suit on the Court at Westminster by challenging two landowners in the county to trial by battle, as though he were himself of knightly class. Maitland drew attention to Hervey's seal depicting a knight with drawn sword on a galloping horse.[1] Seals can give a fair indication of the personal tastes of their owner, especially when they are decorative and pictorial. One would suspect that Hervey resented the fact that he was not a knight by hereditary right and that it was his ambition to attain the appearance and status of that class. Consistent with such ambition would be the indications of an extravagant way of life, hinted by Maitland, and the lavish way in which Hervey endowed the local charities. Generous benefactions were often as much status-symbols as evidences of piety, and Hervey was generous to Barnwell, to the Nuns of St. Radegund, and to the Hospital, to the detriment of his ancestral inheritance, so that he had to purchase or rent other lands to keep up his acreage. These transactions would seem to be the first indications of financial instability in the Dunning estate.[2]

To turn from the social aspect of Hervey's gifts to the topographical; although it is not possible to show exactly where and how much Hervey gave away, about 12 acres (7 in Chesterton) are certainly recorded as going to the Hospital, and about the same, mainly in the Cambridge Field, to the Nuns. Where there are

1   *T. & B.*, pp. 65–6 and p. 165. This seal is an almost exact replica of that of Sir Baldwin Blangernon.

2   Merton, M.1587, 1589.

place-names to locate these lands, the areas of Grithow Field and the St. Neot's Way predominate, and a plotting of the pieces which eventually fell to the Clerks of Merton shows the same concentration on this upper end of the Cambridge Fields all in the vicinity of the gravel ridge of Grithow and at the side of the St. Neot's Road.[1] The obvious exceptions are the "pieces" in Le Dale or Dedale, which can be shown to be leases from the Canons of Barnwell of the land formerly of Ailgar the Noble, *que terra iacet pro triginta acribus* and is in two *culture* namely in "Le Dale" towards Cotes and in Godivesdole (later written Goidesmedole). The topography of the Chesterton properties has not been worked out, but the frequent mention of the ways to Histon, Beche, Oakington, suggests a similar balance in what may, in the distant past, have been a set of arable fields radiating from Castle Hill, with a backbone, rather than a boundary, along the Huntingdon Road. The indications are very faint, but the concentration of what can be plotted sufficient, we think, to suggest strongly that the Dunning inheritance went back to an arrangement of fields based on the natural topography of the upland settlement, rather than on the system using the Huntingdon Road as a boundary that survived from the later Middle Ages into modern times.

We have already commented on the curious arrangement of the boundary here and the interrupted pattern of the Hundreds. We have no means of knowing exactly what ancient patterns of government and lordship were disturbed in the late Saxon period and again at the Norman Conquest in obedience to the dictates of military convenience or necessity. It must suffice to say that two major disturbances took place within a relatively short period, and the general effect was to thrust the bounds of the royal manor of Chesterton right into the ancient hilltop settlement. We know that twenty seven houses were destroyed to make room for the Norman castle, but not what else was destroyed or distorted. It does not seem impossible that an ancient lordship, based on the Castle Hill area, was almost squeezed out of existence by the setting up of the new feudal structure. And if this was so, perhaps its remnants were the inheritance of the House of Dunning. Such a supposition becomes slightly less fanciful in view of the very odd bulge northwards made by the Cambridge-Chesterton boundary as leaves the edge of the Castle to go up the Huntingdon Road. Within that bulge is the croft of about five acres known as the Sale Piece. In this piece, we may have the remains of the homestead of a pre-Conquest lord. As Milner-Gray pointed out, "sael" in Anglo-Saxon means "hall", and the Dunnings were stated to have enjoyed the plot by long seisin from the days of the Conquest.[2] He believed it to have been the original home of the Dunning family, or at least that their first home was adjacent to this five acre croft at the back of the Castle.

The Liber Memorandorum Ecclesie de Bernewelle thus describes the route of those who were perambulating the bounds:—*Incipientes ad locum qui uocatur Armeswerk circuibant fossatum Castri, ascendentes ad locum qui uocatur Aswykston, et descendentes fecerunt transitum per medium curie Scolarium de Mertone per uetus fossatum usque ad riueram.*[3] From which one would conclude

1 Merton, M.1568, 1569 and 1606. In furlongs 19–21 of Middle Field.
2 J. M. Gray, *op. cit*, p. 1.
3 *Bernewelle*, p. 167–8.

that the boundary crossed a courtyard belonging to the Merton property and a courtyard presupposes some sort of farm or house.

The records of Merton College have several deeds relating to the sale or mortgage of the Sale Piece, which apparently was a single *cultura* with its own surrounding ditches, e.g. *v acre iacentes in crofto que vocatur Le Sale ad capud ville Cantebrugg' iuxta castellum cum fossatis dicto crofto pertinentibus.*[1] No house is included in any of these leases, but in a later Merton deed a messuage on which there is rentcharge to Merton is described as extending from the highway to the land of the Clerks of Merton called 'le Sale'.[2]

Although the position of the ancestral Dunning house in the town of Cambridge is not certainly known, its existence is shown by the document of 41 Hen. III, by which Eustace Dunning, son of Hervey, mortgaged the greater of his possessions to Guy of Barnard Castle.[3] In this deed the two capital messuages of the Dunning family are clearly distinguished, as one is specifically included in the mortgage and one specifically excluded. The house which is included is obviously the older house, which has been in the family longer and which is described in the following terms, *totum mesuagium meum quod fuit Hervei patris mei in quo habitavit* (sic) *in villa de Cant.'* The house which is excluded is described in these words, *excepto capitale mesuagio quod emi de Badwino blangernon.* The preceding accompanying documents make it clear that this latter is none other than the famous Stone House, later known as Merton Hall. The deed of purchase, sealed with Blangernon's seal, dates from the time when Henry de Colville was sheriff and was therefore between 1249–1251. Eustace, who had bought the Stone House only seven years or so before he fell into grave financial difficulties, and who was living in it at the time of the mortgage, was at first unwilling to let his new house go. In a subsequent bond to Guy, dated 1260, he had to pledge both his capital messuages but nevertheless managed to stay in the Stone House until his death, and his son was only removed by an action of *mort d'ancestor* brought by Guy's heir ten years later.[4] Such was the tenacity with which the Dunnings held on to their last, most expensive and most desirable of purchases.

We have gone into this point in some detail, as it is often stated in print that Hervey owned the Stone House and lived in it for the greater part of his career. The Merton deed of purchase shows that he only rented it from Blangernon on a twelve-year lease towards the end of his life, and his son Eustace converted this lease into a purchase by paying *triginta quatuor marcas premanibus in gersumam preter viginti duas marcas quas nuper mihi dedit pater predicti Eustachii pro predicto mesuagio habendo ad terminum duodecim annorum.*[5] The customary rent to the Blangernons was to be a pound of cummin in lieu of services. The house was described as, *totum capitale mesuagium quod iacet iuxta*

1 Merton College, M.1577, 1578 and 1579.
2 Merton College, M.1575 and 1618.
3 Merton College, M.1545, preceded by M.1543 and 1544. There is a fair possibility that the other Dunning house is the one later redeemed from Merton by Thomas Dunning. C.C.C. Cartulary of Sir John de Cambridge,.
4 Merton College, M.1548, and Milner Gray, *op. cit.*
5 Merton College, M.1544 and 1545.

mesuagium dicti Eustachii cum hulmo et omnibus edificiis et pertinenciis et cum libertate falde et tauri et omnibus aliis pertinenciis ad predictum mesuagium spectantibus. "The messuage of the said Eustace", mentioned as next to the Stone House, had been granted to Hervey by Baldwin Blangernon some twenty or so years previously, and the deed is in the Merton records. It was quite a small house with a *gersuma* of only two marks and was presumably incorporated in the Stone House estate after Eustace acquired both. It was probably the wing used as a solar, described by Milner-Gray and said to be of contemporary architecture with the main hall. The whole was approached by its own lane, *venellam que itur versus domum Baldwini Blangernon,* and another small house in the lane was occupied by Geoffrey Brun in 1233. Later, the whole was described as a stone house and buildings *cum gardino et tota curia adjacente.*[1]

It is clear, therefore, that the nucleus of the future Merton Manor was purchased from Baldwin Blangernon and had not formed part of the original Dunning estate. This inheritance of Eustace Dunning proves on examination to be nearly as composite as the Merton manor which succeeded it. Very roughly, it might be described as a union of Dunning lands and houses without lordship to Blangernon lordship without lands. This is a very dangerous statement, since it is nowhere specifically stated that among the various rights and appurtenances that went with the Stone House was the right to hold a court for one's tenants. All one can say is that Baldwin Blangernon the elder had been lord of a considerable fee, the greater part of which lay in the West Fields. A tremendous amount of this property went to Barnwell, and accounts for most of the large holding of the Prior of Barnwell there.[2]

The jurisdiction of the Blangernon fee would have been very largely dissipated by the lavish grants to Barnwell, since the Prior of Barnwell, who was already lord of the Manor in Chesterton and elsewhere, would withdraw the suits of his tenants in the West Fields into his own courts. What happened to the residue, and whether it passed to the new owner of the Stone House as part of his rights and appurtenances, may well have been no clearer in the thirteenth century than it is now. The fluid situation with regard to rights of jurisdiction was fertile ground for interminable law-suits, provided one had the means to pursue them. It seems highly probable that Eustace fitz Hervey, had he not fallen into financial straits, would have been able to make good a claim to manorial jurisdiction by the processes of law backed by influence and cash. Many ambitious men, both before and after him, did so successfully. It is interesting in this connection to note that in 1240, just after his father's death, Eustace Dunning attempted to claim the advowson of All Saints by the Castle from the Blangernons. The attempt failed, since this advowson had already been quitclaimed to Barnwell some twenty years previously. But the fact it was made shows that Eustace Dunning was anxious to lose nothing that he considered part of the former Blangernon rights that should now be his.

By purchase then, if not by inheritance, the family of Hervey Dunning had achieved most of the outward trappings of the feudal or manorial lorsdhip they were so anxious to possess. We can even detect, perhaps, a slight shift in emphasis

1 Merton College, M.1574.
2 Maitland's summary of the evidence in the *Rotuli Hundredorum, T. & B.,* pp. 140–152.

from the Chesterton area, where the entrenched position of the Prior of Barnwell blocked any road to extension of rival influence, to the West Fields side, where the situation was more fluid and the collapse of the Blangernons appeared to open opportunities.

The Dunnings and the Blangernons had been old rivals on Castle Hill, and Hervey had gone to law with Baldwin the Elder early in his career. The Dunnings must have been envious of the Blangernons, whose name suggests Norman or French descent and a nearer relationship to the Norman-French military upper classes than the Dunnings could boast, though they gave their sons the fashionable Norman-French Christian names. The Blangernons were obviously richer than the Dunnings, their tallage higher, their gifts to charity more lavish, their house more splendid. When the Blangernon son, of the same name as his father, and no doubt suffering from his father's lavishness, got into financial difficulties, the Dunnings must have felt that it was their good fortune to be able to step into Blangernon shoes, the moment the family had been waiting for. First by lease, and then by purchase, they edged their way into the Stone House.

Incidentally, Hervey's brother Adam (as will be seen when the Mortimer estate is discussed) was at the same time farming half a knight's fee with suit of court and attendant trappings from the Mortimers, whose estate was based on Newnham. Both brothers, one would suspect, were at the same game; buying or renting from the feudal aristocracy (whether of ancient lineage in decline or arriviste does not here concern us) the positions they hoped in due course to make their own. They were proud and ambitious men.

What they obviously did not realise was that the price they were having to pay, namely the burdens they were laying on their estates, was far greater than they could bear. It was almost certainly the strain of having to pay for their position, not only in initial purchase, but in recurrent rents, dues, taxes, aids and feudal charges, as well as in the necessary expenditure to keep up the ostentatious style of living that was expected of them, which brought Eustace fitz-Hervey, as Maitland guessed, "into debts and mortgages". For the ready cash of which he was always desperately short, he was prepared to alienate the long-term sources of his family wealth, his customary dues and rents. Such a course could only lead to disaster, and that rapidly.[1]

There is ample evidence for the later Middle Ages, and in some areas for as early as the thirteenth century, that the hereditary land-owning classes were finding it difficult to maintain their wealth if they lived off the land alone, the "view, for which there is greater evidence, that land was a wasting asset", i.e. fixed rents and customary dues remained the same, some might even be lost, while other costs gradually rose.[2] The genuine military aristocracy could to some extent redress this situation, if they were lucky in having healthy sons, by mercenary service, the spoils of war, or a good marriage to an heiress. But a land-owner who was below the knightly class, a pseudo-feudal lord, unless he had alternative sources of income such as might be afforded by trade, or had

1 Maitland and Milner Gray both outline his career. Various Merton deeds confirm the details, e.g. M.1548, M.1555, M.1575-8.
2 K. B. Macfarlane, *The Nobility of Later Medieval England,* Ford Lectures for 1957, Oxford, p. 14.

a clever member of the family 'on the make' in legal or clerical circles, was almost bound to go under. It is interesting to note that the only members of the Dunning family who survived into and flourished in the fourteenth century, were those who married the daughters of tradesmen of the town, Thomas Dunning and John Dunning. John's daughter and heiress, Joan, with her marriage brought lands and respectablilty to that parvenu lawyer and property speculator, John de Cambridge.[1] To quote Macfarlane once more, "Those with the most unquenchable appetite for land were most often the men who had been born without much of it."[2] With the de Cambridges we come to a new type of owner who appears frequently in the terrier of the West Fields, Richard le Tableter, Roger de Harleston, the two Morices, for instance, whose names would have been unknown a century earlier. These owners of large polyglot holdings are buying and selling land as a steady if clumsy, way of converting the capital they have acquired elsewhere into income, with the possibility of realising it again when necessary. With them we might include Geoffrey Seman who, although coming from a local family (Seman of Newnham), is found in the early fourteenth century manipulating property transactions and cash loans on a much wider scale, comparable with the way in which Sir John de Cambridge made his fortune. Evidence for this comes from two documents of 1328 and 1342, both transactions carried out in London, one involving a manor as far afield as Somerset. Strips in the fields could often change hands as incidentals to a sale of urban or family property and acquisition in this way may help to account for the haphazard distribution of the lands of Corpus Christi College.[3]

These then, are the phases of ownership we can see in the West Fields in the century and a half preceding the writing of the Terrier. In 1200 the last of the older feudal owners are fighting for survival against ambitious local families who hope to succeed them. By 1260 those same local families are themselves going down under a rising tide of wealth which has no basis in the land. But waiting in the wings are the last and greatest threat to private owners old and new, the corporations who never die, and which are not even destined to be dissolved at the Reformation, the Colleges. Slowly and inexorably, they are to creep over the West Fields.

Lordship in the West Fields

B. Mortimer and St. John's Hospital

In the court case of 1803, Merton College v. The Corporation of Cambridge, it was noted that St. John's College, the successor to St. John's Hospital, had similar rights of faldage and access over the West Fields to those enjoyed by Merton College. The provenance of these rights should be now discussed, together with those of a third College, Gonville & Caius, which never in fact entered the arena in 1803. Had it been the good fortune of Gonville Hall to

1 C.C.C. Cartulary of Sir John de Cambridge.
2 K. B. Macfarlane, *op. cit.*, p. 83.
3 C.C.C. Misc. B. 66 and B. 89.

acquire the Manor of Mortimers in Newnham in the Middle Ages, rather than after 1500, there is little doubt that Gonville & Caius College would have been in a strong position to join battle with Merton and St. John's as claimants to ancient lordship in the West Fields. But by the time the College received this manor, in 1506, at the hands of the executors of the Lady Anne Scroope, last heiress of the Mortimer family, the Corporation of Cambridge was riding high on a theory that set it up as a mesne lord immediately beneath the Crown and above any tenant paying haw-gafol.[1] Historically the theory was ridiculous in that haw-gafol, a tax on ancient town properties, predates the existence of the Corporation by several centuries. Moreover in the case of Mortimer's manor, the haw-gafol of 18s. was paid on scattered dwellings here and there in Newnham and Cambridge which happened to belong to the family, and not on the manor-house, mill and adjacent land, which were the nucleus of the Mortimer manor. In any case 18s. was a small sum to set against the ancient rent of £4 of the Third Penny of the Earls, originally granted by Earl David, King of Scotland, to Sir Robert de Mortimer, to be paid annually by the Mayor and bailiffs, in recognition of the fact that they enjoyed the use of the Tolbooth, formerly the House of Benjamin the Jew, and anciently part of the donation of the Crown to the Earl David and thence to Mortimer.[2]

Details of this £4 rent appear in the Mortimer rentals of the fifteenth century, *In primis de Maiore et Ballivis ville Cantebrig' pro quodam mesuagio iacente in foro cum prisona et shoppis eidem annexis quod vocatur le tolleboth quod quidem mesuagium reddet per annum ad festa pasche et sancti michaelis equis porcionibus manerio predicto videlicet quatuor libras argent' de quibus quidem quatuor libris argenti resolvi debent maiori et ballivis. . . xviii s. pro haugafol quolibet anno. Et si dictus redditus quatuor librarum argenti a retro esse contigerit vel non solvitur tunc bene liceat domino manerii predicti vel eiusdem assignatis in molendinum aquaticum domini regis ville Cantebrig' ingredi et petra molendinum pusillum ac omnia alia necessaria in eodem molendino manucepere abducere retinere sine aliqua contradictione quousque de predicto redditu et arreragio plenarie fuerit satisfactio.*[3]

But the Corporation of 1500 was in no mood to admit that it held anything of anyone lower than the Crown, least of all its Guildhall. A series of hard bargains were driven with the Lady Anne, who suffered from the double disadvantage of being an absentee owner and a woman. By one, the £4 rent (regarded as a sort of rent-charge) was extinguished, by the other, the Newnham Mill was turned into a subsidiary of the King's mill, run by the same milling tenant appointed by the Corporation. This made it impossible for the Mortimers to seek redress according to ancient custom by sending their man to impound the stones of the King's Mill. The Corporation moreover had taken out a long lease of all the Mortimer lands, so that in course of time it was almost impossible to see who was tenant of whom in respect of what. One suspects that some of the

1   *T. & B.*, pp. 186–7. The name Scroope is preserved in Scroope Terrace, Trumpington Street.
2   See charter of Sir William Mortimer in Appendix B and *Rotuli Hundredorum*, ii, p. 356, and H. P. Stokes, *Studies in Anglo-Jewish History*, pp. 136 and 248.
3   G. & C.C. *XIII*. no. 23.

ancient rights advanced by the Corporation in support of their claims to lordship in later centuries came originally from the customary rights of the Mortimer Manor. Robert Brady, Master of Gonville & Caius College writing a century later, recognised that the position of the College vis-à-vis the Corporation of Cambridge had deteriorated during the time when the Corporation held the Mortimer manor on a 99 year lease:[1]

> "By reason of this long lease the college lost almost all their quit rents, a sheep walk, and a free bull and boar, and about 40 acres of land, by changing their marks, bounds, etc."

The small college of Gonville Hall seems to have been quite ignorant of the true state of the case and to have set the seal on the Corporation's pretensions by applying to it for a licence of Mortmain to receive the Manor of Newnham. This licence is referred to by Maitland as "the high-water mark" of the theory just described and he quotes from it the assertion that the Mortimer manor was "holden of the mayor, bailiffs and burgesses *as of their high gable* by the yearly rent of 18s.[2] The original licence is still in the Treasury of Gonville & Caius College, sealed with a magnificent seal showing angels bearing aloft the badge of the Corporation.[3] After this, legally speaking, the College had no leg to stand upon and apart from the payment of a fixed rental "for the Newnham Mill", were in subjection to their tenants, the Mayor and Bailiffs of Cambridge. At the time of the Corpus Terrier, however, the position of Mortimer looks very different, and supporting documentary evidence confirms this impression.

The earliest terriers we have of the Manor of Mortimers are in fifteenth century copies of what seems to have been an original terrier of the late fourteenth century, one generation after the Corpus Terrier, to judge from the names of the quondam owners quoted on it. From this we find that the headquarters of the Mortimer Manor was a house in Newnham which stood somewhere in the gardens of the present Newnham House and Ashton House, or possibly just in the Caius Fellows' Garden. (The house appears on Hammond's plan of 1592 but has vanished by the time of the late eighteenth century maps). On the road frontage of this manor house were a number of tenements all paying rent-charges to Mortimers and obviously built in front of the manor-house on the Manor land. One of these was called Nightingale's, after a former tenant, and another was called "Cuckoo's" or "Cukkous" (might this be a medieval joke?). This little piece of ground, bounded by Froshlake Way, was the only built-up ground on the west side of the High Street. The other built-up area on the east side was in a triangle with the mill at its apex, and these houses also paid rents to Mortimer. Even the main houses of St. John's Hospital and

---

1 G. & C.C. MS 617. Quoted by J. Venn. *Biographical History* Vol. III, p. 69.
2 *T. & B.*, p. 85.
3 G. & C.C. *XIII*. Gonville Hall does not appear to have been very much on the spot in more general matters of estate management, if we may judge from a note on the flyleaf of their surviving Account Book. "In yeare of ower lord god 1534 on cockeman fermour of ower mylle at newnham made a new mylle of his own proper costis and chargis without ower consent nor havyng theroff ony knowlege".

Corpus Christi College, which they subsequently used as the "home-farms" of their Newnham-based tenants (replaced by Newnham House and Ashton House in the early 19th century), were on Mortimer land and paid these rents. A small outlying portion of the parish of St. Peter's (later St. Mary-the-less) surrounds this built-up area, while the mill and its adjacent tenements falls into an outlier of St. Botolph's. Here we find the nucleus of the Mortimer Manor, the manor house, mill, and demesne land, partly built upon and partly in the adjacent meadow of Mortimer's Close (later Butcher's Close) and Mortimer's Meadow. This is possibly the only part of the West Fields never to have been under the plough, and here the trees still grow more luxuriantly and to greater height than on the former arable of the Carmefield.

The manor house of the Lord Robert Mortimer, as it is called in the terriers, is the subject of an interesting series of deeds in the Caius College Treasury.[1] By the first of these, dated 1311, the Prior of the Carmelite Order in Cambridge agrees, reluctantly, to release to Sir Constantine de Mortimer the house formerly held by the Carmelites in Newnham, and this decision is confirmed by the Provincial Chapter at Lincoln in August of the same year. The Provincial Chapter state that they are acting "in the interests of that peace enjoined upon them by the apostle" but one suspects that they had a poor legal case for trying to hang on to the house. Sir Constantine subsequently leases the house, "formerly held by the Carmelites", to his son, William, for life. Later deeds show the house in the Mortimer family, normally granted to the eldest son of the family along with the rest of the Mortimer Cambridge property, prior to his succession to the main family lands, and this remains the case until the mid-fifteenth century, when the "house called Mortimers" is let to various other tenants, usually of good class. The proved association of the house with the Carmelites is a direct link with the large block of land called the Carmedole immediately behind the manor house and its close. This is surely ancient demesne, receiving its name when the Carmelites lived in the house and worked upon it. So to the house, mill, meadow and closes we add the Carmedole and there, on the map, is quite a respectable manor.

The statement in the Hundred Rolls that the Carmelites had their Newnham house from Michael Malherbe does not conflict with its being, in origin, a Mortimer house. Malherbe, whose main possessions lay in the corn-trading area of St. Michael's parish, may well have rented the house for its proximity to the Newnham Mill. It is noted that there is some doubt as to whether his grant to the Brothers had been properly confirmed.[2]

How much the Friars extended and adapted their house during their forty years of residence, as described in the Barnwell Book,[3] is a matter of conjecture. The Close Rolls refer to gift of twelve pairs of rafters, not enough to indicate a very large structure, but enough to make the Brethren reluctant to let it go to the Mortimers. The documents do, however, show clearly that the Carmelites were home and dry on about the only bit of Newnham well above the 25ft. contour line. The traditional picture of the Carmelites "on an island surrounded by fen and

1 G. & C.C. *XIII* nos. 3–6, sqq.
2 *Rotuli Hundredorum,* ii, pp. 360–1.
3 *Bernewelle,* p. 211.

osier beds intersected by small streams" appears to be a piece of imaginative fiction.[1]

One might have expected that when the Carmelites crossed the river in 1292 the blocks of land which they had formerly worked and to which they had given their name would have reverted to the Mortimer manor with their house. Surprisingly they are found belonging to the Hospital. It has been suggested, though no documentary proof has been forthcoming, that with the removal of the Carmelites over the river, some exchange or arrangement was made by which the Hospital supplied land for the Carmelites to work nearer their new home, taking in exchange the Mortimer arable in the transpontine fields. In connection with this suggestion it is pointed out that the Mortimer Manor had several large "doles" in the vicinity of the Trumpington Road in the later Middle Ages. These however, could equally well have a different and more ancient connection with the Mortimer manor and its antecedents. The Carmedole is the most striking but not the only instance of land which might reasonably be expected to belong to Mortimer but which in fact belonged to the Hospital. (e.g. One could point to a number of small plots, balks, headlands which look like ancient waste bordering the oldest tillage in the Newnham area, all to St. John's.) While the map alone makes it clear that the Mortimer/St. John's Hospital connection needs more thorough investigation, other documentary evidence supports the view that they are ancient partners. There is the odd fact that both manorial-type courts held in Newnham in the thirteenth and fourteenth centuries, to which Maitland refers,[2] were held from adjacent houses; that of St. John's being held in one of those houses which, from its ground rents, appears to have been built on the former demesne of the other. The most logical explanation is that here we have a split manor, stemming from a single original owner, who at some point gave half of his holding to the Hospital, and with whose descendants the Hospital did some minor exchanges when it became its policy to consolidate its Cambridge property in the West Fields.[3]

Such an original single owner is not far to seek, namely the first Sir Robert de Mortimer, who served and was rewarded by King John.[4] We know from the Hundred Rolls that he gave the largest single foundation endowment to the new Hospital early in John's reign and that it amounted to a whole carucate (c.120 acres).[5] The same Sir Robert, or his son, made an arrangement with Adam Dunning, brother of Hervey fitz-Eustace, to farm his remaining Cambridge properties in return for a fixed rent. Details of the arrangement are given in extant charters and confirmed by the entry in the Hundred Rolls. The only slight difficulty is that the carucate given to the Hospital is said to have come from the Crown, while the other property, amounting to almost a carucate, is shown by charters to have come from the Earl David of Scotland. But as the

1 *VCH Cambs.* ii, p. 282.
2 *T. & B.,* p. 183.
3 *T. & B.,* p. 183 n. gives an example of such an exchange.
4 This is not the place to go into the antecedents of Sir Robert de Mortimer, but it seems that as far as Cambridge is concerned, he is a newcomer who has received his property as a reward for military service at the end of the twelfth century.
5 *Rotuli Hundredorum,* ii, p. 359.

Earl had himself only recently received his Earldom of Huntingdon, and had had it confirmed to him along with some other newly acquired lands, by the Crown, it might seem reasonable, in a summary of title, to refer to the whole as having "come from King John", as the Hundred Rolls does. (Similarly in the reference to the Tolbooth, or House of Benjamin, for instance, the whole descent from the forfeiture to the Crown, through the Earl David to Mortimer, and thence to the mayor and bailiffs, does not appear as fully as in the extant charters. And again, sums about the Earl's Third Penny, based on the Hundred Rolls alone, do not add up to correct totals.)

The earliest extant charter of the Mortimer family relating to this land is that of Sir William, son and heir of Sir Robert de Mortimer.[1] By it he grants "all my land, except the house which was Benjamin's, which I held of the Earl David in Cambridge, Newnham and Barnwell, with demesne, service, mill rents etc. with the four pounds of the third penny of the Earl David of the town of Cambridge... in fee and heredity", for the annual farm rent of thirty-two marks. The farmer is Adam fitz-Eustace, brother of Hervey Dunning the Mayor, whose desire to adopt the life-style of the feudal aristocracy has already been described. By the next charter, the farm is confirmed to Leonius Dunning, son of Adam, and it is he who is found holding it in the survey of 1243 and in the Hundred Rolls.[2] It is assessed as half a knight's fee, and is held of Robert le Bruce, successor of the Earl David. The descent of the half-fee is set out, making it clear that it is the same property, held under the same conditions, as the original grant of Sir William de Mortimer, quoted above. The arrangement with the Dunnings continues for two or more generations, to 1337, when John, son of Robert, son of Richard Dunning, son of Leonius, returns the whole property, rents and rights, to Thomas, son of Sir Constantine de Mortimer, son of Sir William. The full details of the Mortimer lands or "manor" as it was in 1311 are set out in the Charter of Richard Dunning to Robert Dunning.[3]

Before the arrangements with the Dunnings, eight acres of Mortimer land, mostly in very small parcels, had been passed to the burgess Thomas Tuillet. These eight acres have an interesting history which illustrates how small parcels of land belonging to a large fee could subsequently be lost to that fee. There is a copy in the Cartulary of Sir John de Cambridge of the original deed of grant, dating from the first half of the thirteenth century, and it records a rent-charge to Mortimer of 4/6d. per annum in lieu of services.[4] The transaction is made the subject of a court action by Barnwell Priory when the right of Robert Huberd, heir of Thomas Tuillet, is challenged.[5] But Robert was found to be the legitimate heir, and soon afterwards appears to have passed these acres to Bartholomew Goggying, one of the leading townsmen. Just to make sure of his title, Goggying obtains a confirmation of his right in the eight acres from Leonius Dunning, then farming the Mortimer fee, and the price he pays for this is to have his rent-charge put up to 5/6d. per annum. This deed is dated 18 Ed. I (i.e. 1290).[6] The eight acres, made up of a number of small strips, then pass by inheritance in

1, 2, & 3. Mortimer charters from G. & C.C. muniments, quoted in Appendix B.
4   C.C.C. Cartulary of Sir John de Cambridge, fol. 33.
5   *Bernewelle,* p. 115–7.
6   C.C.C., *VII.* 6–8, 38 and 65.

the female line to the de Cambridges, and thence to Corpus Christi College. But by the time they reach the College their connection with the Mortimer fee is no longer recorded and the 5/6d. has disappeared.

It is possible that Sir John de Cambridge extinguished the rent-charge for a cash payment but if so, the record of the transaction has perished. In a parallel case, also documented by Corpus charters, Baldwin Blangernon disposes of a similar holding of about the same total acreage, made up of several small parcels. The rent-charge in lieu of services is half a pound of cinnamon. The holding is again fragmented, different parts passing to different families and one wonders who was left with the cinnamon obligation. But finally six of the original parcels come into the hands of the de Cambridges, together with some other ex-Blangernon property. In 1322 Sir John de Cambridge buys a quittance of an old rent of one lb. of cinnamon from the heirs of Baldwin Blangernon, and the clue which makes it possible to trace this particular holding vanishes.

To the same sort of loss, through fragmentation and subinfeudation, we might perhaps attribute the fact that nearly all the Mortimer lands that survive into later times are the large blocks—a point made by Maitland.[1] They are, in fact, nearly all "doles" and there are very few arable strips of moderate size scattered in a normal pattern. It is particularly noticeable that apart from the Mortimer manorhouse at Newnham, the Close and the Newnham Mill, there is no Mortimer land in the Newnham area. Yet there are large holdings round Castle Hill and at the top end of Little Field, and other large pieces in Middle Field and in the area of the Trumpington Road. (See table below). Allowing for losses, such as were common under all but the most vigilant of owners when subinfeudation ran unchecked, the total Mortimer acreages comes to 99 (as assessed in later documents) + 8 (acreage to Tuillet) + (say) c.10 in small lost pieces, and is thus roughly a carucate—equal to what was given to the Hospital.

This again leads to the conclusion that what the Earl David received and passed to Sir Robert de Mortimer was probably itself a composite holding, made up to be the equivalent of a knight's fee (i.e. roughly two carucates) of three main basic constituents; firstly a holding round Castle Hill suitable for a knight of the Shire and a military tenant, (we are very much in the dark about the holdings of the men who did Castle duty in the really early period but there is considerable support for the view that some holdings near the Castle were originally allocated in connection with military duties); secondly, a Newnham-based holding of which the ancient mill was the centre and raison-d'etre, stemming perhaps from very ancient riparian and water-rights.[2] For the Domesday owner of the Newnham Mill, the Count of Brittany, like his predecessor Edith the Fair, also held the Grantchester Mill and it is in Grantchester parish that the cut to Newnham Mill begins. (The holdings on the other side of the river may even have a connection with water rights but this is not a matter relevant to the West Fields. It is significant, however, that the Mortimer terriers consider the property in Newnham first, then in the Trumpington Road area, then in Barnwell, and finally round Castle Hill. It was perhaps a matter of convenience, arranged with the Hospital, that the Mortimer lands

1   *T. & B.*, p. 179.
2   *T. & B.*, p. 180 and 180n.

Showing the realignment of the Roman (Barton) Way, and the ownership in large blocks, much to the Hospital and to Mortimer.

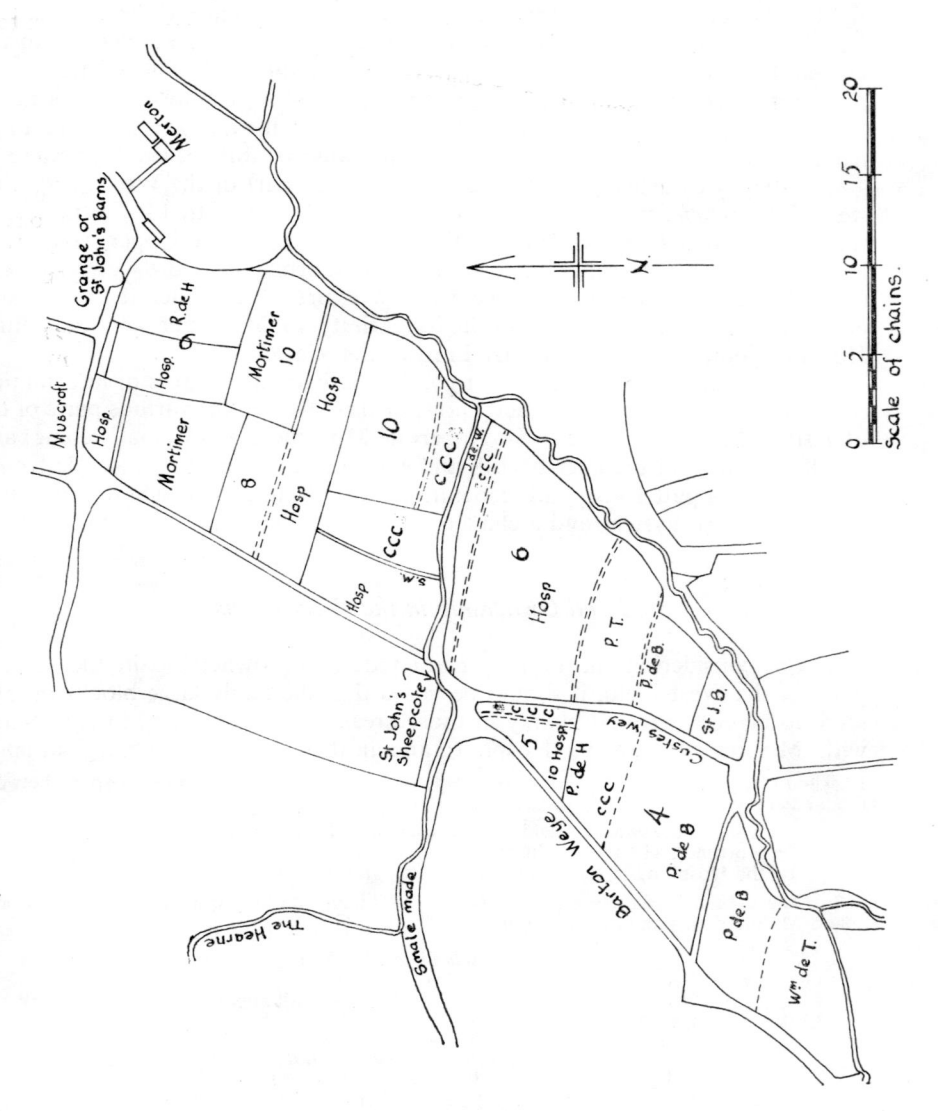

in the last three areas remain extensive, while that round Newnham seems to have been given to the Hospital.)

Finally, there is the enormous piece of Mortimer land, the 41 selions lying right in the centre of the Middle Field, on the slopes of what is called Alderman's Hill. Next to it on the south and west are two large blocks belonging to the Hospital, these three blocks between them contain the whole of this small promontory commanding the Barton Way approach to Cambridge. This surely has some significance, when taken in conjunction with the fact that this area, from the evidence of older field nomenclature, was possibly the last big area to be systematically assarted from the waste. This is the part of the West Fields that we believe to have remained "no man's land", in the sense that it was neither in the Newnham field system nor in the arable fields of the Upper Town, but remained 'moor' ground, perhaps where the Ealdorman once drew up his troops for the defence of the town by the western approaches. When it finally came under the plough, it fell almost in its entirety to Mortimer and St. John's Hospital, joint heirs of the former Lord of the Waste.[1]

Further back than this we cannot go, for there are no sources of information sufficiently detailed to show exactly how, or from whom, the various parts of the donation of the Earl David to Sir Robert de Mortimer had originally come; and the disturbances of the later twelfth and early thirteenth century have destroyed the Domesday pattern beyond recognition. All this, like the origin of Newnham itself remains a mystery and a challenge for further study.

Land Ownership in the West Fields

Before considering the more general pattern of ownership in the various parts of the West Field, it should be noted that the really large blocks of land seem to be concentrated mainly in those areas already indicated in connection with Mortimer/St. John's Hospital; and in the 'doles' of the marginal lands

[1] The most striking instances of areas in which Mortimer and St. John's Hospital between them have a virtual monopoly of the arable in large blocks are

1. The area of the Middle Field on Alderman Hill (Qs. 16 and 17)
    To Mortimer, 41 (now 38) in one piece
    To the Hospital, 45 (now 42) in two pieces (adjacent to Mortimer).

2. The area immediately adjacent to Castle Hill, in the first quarentenas of Grithow and Middle Fields and at the top of Little Field.

		No. of selions to Mortimer
Grithow	Q.1	4+4+8 (Chalkwell)
Middle	Q.1	4+3+2 (In Muscroft, all alternating with Hospital)
Little	Qs. 8, 9 & 10	12+11

		No. of selions to the Hospital
Grithow	Q.1	8+10 (super le Cleypittes)
Middle	Q.1 Muscroft	5+4+5 (all alternating with Mortimer)
Little	Qs. 8, 9 & 10	14+3+7+16

3. The area of the Carmedole. Here the close on which the manor-house stands is to Mortimer and so is the meadow, later Butcher's Close. But the arable is all to St. John's in the following large pieces

Carmefelde, Qs. 13 & 14   20 (Carmedole) and 6 (Whelp's Acre) and in various small pieces, such as the headlands in Qs. 1 and 6.

(i.e. both those which are marginal topographically, at the far edges of the Fields, and those which are 'marginal' in the normal agricultural sense). Some of these bear names of exceptional interest and deserve special mention.

Priorsdole, through which ran Grithow Way and marginal on account of its awkward slope, consists of selions of St. John's Hospital. But who was the Prior from which it took its name? Neither the Prior of Huntingdon nor the Prior of Barnwell was in the habit of letting anything go in this area, and seem unlikely candidates. The only other Prior in the running would seem to be the Prior of the Carmelites, and when it is recalled that the Carmelites had their house on Castle Hill before moving to Newnham, it seems possible that this was their first territory, exchanged for the Carmedole after their moving. This is pure guesswork and documentary evidence on the subject would be welcome.

Erlesdole, in the far S.W. of the Fields and on the Coton boundary between Edwin's Ditch and the ancient Waybalk belongs by the time of the Corpus Terrier to the Manor of Burwash. The descent of this manor, which sprawls untidily "athwart the village geography" as Maitland would say, from Grantchester into Coton and with odd outliers beyond, can be traced through the Duke of Lancaster and Leicester to John Lacy, first Earl of Lincoln by right of his wife (d.1224). The "Earl" of Erlesdole would therefore be the Earl of Lincoln and the period at which the dole was allocated and probably first assarted to the arable was possibly the early thirteenth century.

East of Erlesdole, and also in Deddale, comes the land of the Clerks of Merton, part of which is known in the early Merton deeds as Godivesdole. It lies, say the documents, in two *culture,* on either side of Edwinsditch. Its location so far from the rest of the Merton lands (which do not otherwise come S. of the Madingley Road) would present a problem, had we not the documentary evidence of its descent from "Ailgar the Noble and his wife", who gave it to Barnwell. It was then the subject of an exchange between Barnwell and Dunning the early thirteenth century. It is estimated as being of thirty acres, the ancient virgate, and is the only exact virgate we have been able to detect in the West Fields of which we have the provenance going back into early feudal times. It would be so nice to know whether Ailgar's wife was indeed called Godiva, and when they lived. As it is, we can only postulate that the area was allocated and ploughed up by about the end of the twelfth century.

These large doles are special cases, however, and for a comparison of the more normal patterns of holding in the West Fields, five main areas have been selected. The first of these is the part of Grithow Field towards (though not immediately adjacent to) the Upper settlement, and probably an area of ancient cultivation for the inhabitants of Cambridge-on-the-Hill. (Area I). The second is the far West of the fields, on the Girton, Madingley and Coton boundaries, on either side ot the Madingley Road, and includes the area where the older boundary mark appears to have been overstepped. (Area II). For comparison a section of Middlefield nearer to the river is also given. (Area III). An analysis of Little Field is Area IV. The last area is those parts of the Carmefield on the gravel terrace nearest to Newnham hamlet, which like the Grithow appears to have been an area of ancient cultivation. (Area V).

In all the tables the 'quondam' owner is given where known, and this is particularly important in the case of Corpus Christi College, which represents the former holdings of three important families, de Cambridge, Seman and de Tangmere (or Tanglemere). Another parvenu owner is Roger de Harleston, while the Morys or Morice family, in its two branches, has gathered up far more than their fathers had. From the quondam owners recorded in the Corpus Terrier (and we suspect that a 'quondam' owner is only recorded in this terrier where he has replaced one shown in the original Barnwell Tithe Book and so is relatively recent), we can see the possessions held by town families being concentrated on fewer of their number, if they are not dispersed by gifts to charitable bodies. We see the families in the hamlets adjacent to the Fields' boundaries, Howes and Cotes, gradually being squeezed out by bigger owners.

In the Middle Field, ownership is in the main very fragmented, so that the blocks; Aldermanhill, Brunnforthedole, and an unnamed block of the Prior of Barnwell in Sheepcote Way (Q.15), stand out in marked contrast to the rest. Even the ubiquitous Hospital, Corpus Christi College and Morys family do not do particularly well. It may be noted that the owners primarily holding in the northern Field; the Prior of Huntingdon, Robert Brigham and Roger de Harleston (as heir of Wm. Warde) do not hold south of Sheepcote Way. The owners connected with Newnham and the Carmefield in general; John Pilet, Nicholas Crocheman and Richard Martyn, hold only in the two furlongs on the western perimeter of the area. The University, unrepresented in Grithow Field, holds in the central area of the old Portefield. Any strips in this area of which we have been able to trace the provenance have passed through several hands, mostly of various town families, and this is probably the area which has been subject to the greatest fluctuations and fragmentation in ownership. For instance, John de Comberton is given as the quondam owner for three different current owners in the central area, and he in his turn came from a family which acquired its property piecemeal from other town families.

The owners in Little Field form a cross-section of the usual owners of the Portfield and the Carmefield, from the Prior of Huntingdon to Nicholas Crocheman, and include Thomas de Audele who holds most strongly in the far West, and the University, represented mainly in the southeast. It is noticeable, however, that the owners holding primarily the northern and central areas (the two Priors and Robert Tuillet) are concentrated in the furlongs lying north of Edwin's Ditch and are unrepresented south of it. Their holdings tend to be large, perhaps indicating assarting in largish doles as the Middle Field cultivation was extended from both ends. The Newnham-based owners, on the other hand, (Barker, Crocheman and Pilet, with the Morys family) are only represented in the furlongs south of Edwin's Ditch. Where it has been possible to locate from the Corpus muniments a Little Field holding (e.g. described in early charters as *in campis de Newenham super le broc*) such holdings have turned out to be in furlongs 1–3, perhaps an indication that the area south of Edwin's Ditch was early connected with the Newnham Fields, just as the last four furlongs of Little Field seem to have been early connected with Castle Hill. (See Area III and table of Mortimer). Little Field would therefore appear to be a composite field with its own history, a microcosm of that of the West Fields as a whole.

The owners in the last tabulated area (Area V) of the Carmefield are of particular interest, since extant terriers and charters show that many of them had their homes in Newnham. Some of them hold only in the parts of the Carmefield closest to the village, and here is the nearest we have to ancient "peasant-type" holdings. Some, even if domiciled in Newnham, hold on a rather more extensive scale, like Nicholas Crocheman and Richard Martyn, who each have gathered up the lands of several quondam owners; but none further north than Shepecote Way. Even Corpus Christi College has its Newnham lands from different quondam owners, John de Poplynton and Henry de Tang(le)mere, who do not appear in the Northern areas. In Newnham, Grantchester based families (Richard de Arderne and the Bolour family) appear as owners or quondam owners, and charters of the thirteenth century relating to Newnham lands seem to show that there were more owners with Grantchester connections a century earlier. Most notable among the Grantchester owners are:— William de Sengham, Lord of one of the two main Grantchester Manors; who c.1295 granted a tenement in Newnham to Henry le Pestur, who lived in the isolated croft on Old Newnham Way;[1] Emma de Chalers of Grantchester, who granted along with a messuage in Newnham village, some pieces of land described variously as in Grantchester parish, in Newnham Crofts, in the Field called Olde Newenham, "near the path to the Suthe Milne",[2] and Everard fil. Mariota de Grantesete, also known as Everard de Niwenham, who paid 2/- for view of frankpledge to the Court held at Grantchester[3] and who appears in a charter of 1219.[4]

When we turn to the holdings of religious corporations, we find that the various charities and Religious Houses do not hold evenly over the whole of the fields. Even the ubiquitous Hospital and Corpus Christi College have their relatively stronger and weaker areas. Very noticeably, the Houses known to have been endowed by the families of early feudal owners, the Prior of Huntingdon and the Nuns of Beche, are strong in the areas of older cultivation. The Prior of Huntingdon, successor to the de Troubelvilles, is entirely confined to the northern parts. The Nuns are strong here also, but have another separate patch in Newnham, perhaps from a completely different donor.

The Nuns of St. Radegund and the Prior of Barnwell, both representing slightly later foundations, are strong where their rival houses are weak, namely in the outfield to the far west, where one suspects that assarts were being made about the time of their foundation. Fashions in benefactions, like fashions in names or dress, can be observed in the Middle Ages and a popular new house or charity was usually launched with a rush of donations. So we find the Prior of Barnwell, unable to get a foot in in the older cultivated areas of the West Field, heading the table for the far western areas with 125 selions in pieces of all shapes and sizes. It would be interesting to know how much of this came from Baldwin Blangernon, from whom the Priory received the advowson or All Saints by the Castle and generous gifts in land when he was virtually "lord" of Castle Hill (1219).

1 C.C.C. *XIV*. 4.
2 C.C.C. *XIV* 12, 14 & 31.
3 *Bernewelle,* p. 277.
4 C.C.C. *VII.* 1.

The University, late on the scene, had its endowments from Nigel de Thornton, an early chaplain. They are mainly in the central area of the Portefield, and the further 'outfield' portions of the Carmefield. Unfortunately, the original deeds of these lands, catalogued on the list of University muniments of 1420, have subsequently perished and we do not know their provenance. They may well represent the holdings of an ancient town family, for all tithe to the Round Church.

The outstanding recipient of charity in the Carmefield is the Chantry of the Blessed Virgin in St. Peter's. This is almost entirely the endowment of Sabina née Brithnoth, widow of the burgess John de Aylesham, who appear together in the Hundred Rolls as considerable owners in the parish of St. Peter-without-Trumpington Gate.[1] There is reason to believe that the extensive benefactions made to this chantry were the reason for its absorption into the new building of 1352 and the subsequent change of dedication of the church. It was already known as the church of St. Mary-the-Less at the time of the writing of the Corpus Terrier, but the writer does not appear to have been aware of it; perhaps both names were still current. The tithes of this church, having been appropriated to the new College or House of St. Peter (Domus Sancti Petri), appear under the letters D.S.P.

Tithe Ownership in the West Fields

While it is true to say, as Maitland did, that roughly speaking the transpontine parishes paid tithe in the West Fields and the cispontine in the east, this is something of an oversimplification. Certainly the tithe-owners represented by the abbreviations RAD and EGID between them carry off the lion's share of the West Field tithes. And here it is as well to realise that tithe-owners frequently 'wear two hats', that is, they can be considered either under the title of the institution which has appropriated the tithe, or under the title of the church whose tithes have been appropriated.

Under 'RAD', then, we expect to find the tithes of the two churches belonging to the Nuns of St. Radegund; St. Clement's, appropriated in c.1218, and All Saints in Jewry, appropriated in 1180, both still Jesus College livings. For if these two churches do not come into the tithing system under this title, no place exists for them either in the East Fields or in the West.

Under 'EGID' we find the tithes of the transpontine parishes in fact enjoyed by the Priory of Barnwell. St. Giles' Church had its tithes appropriated for the support of the Infirmary by Prior William (1189–1196). All Saints-by-the-Castle, whose advowson was given to the Barnwell Canons in 1218, had its tithes appropriated in c. 1257–64. After the depopulation caused by the Black Death, the two parishes were formally united in 1365. It is likely that this union and consequent rationalisation of tithing in the transpontine parishes was the reason for the writing of the Corpus (Tithe) Terrier, and possibly a writing or re-writing of its prototype, the Barnwell Tithe Books. St. Peters "ultra pontem" would appear to have lost its identity as a separate tithe-receiver at an early date, though it continued to exist as a separate civil parish until modern times.

1   C. P. Hall, In Search of Sabina. *Proc. Camb. Ant. Soc.* LXV (1974), p. 67.

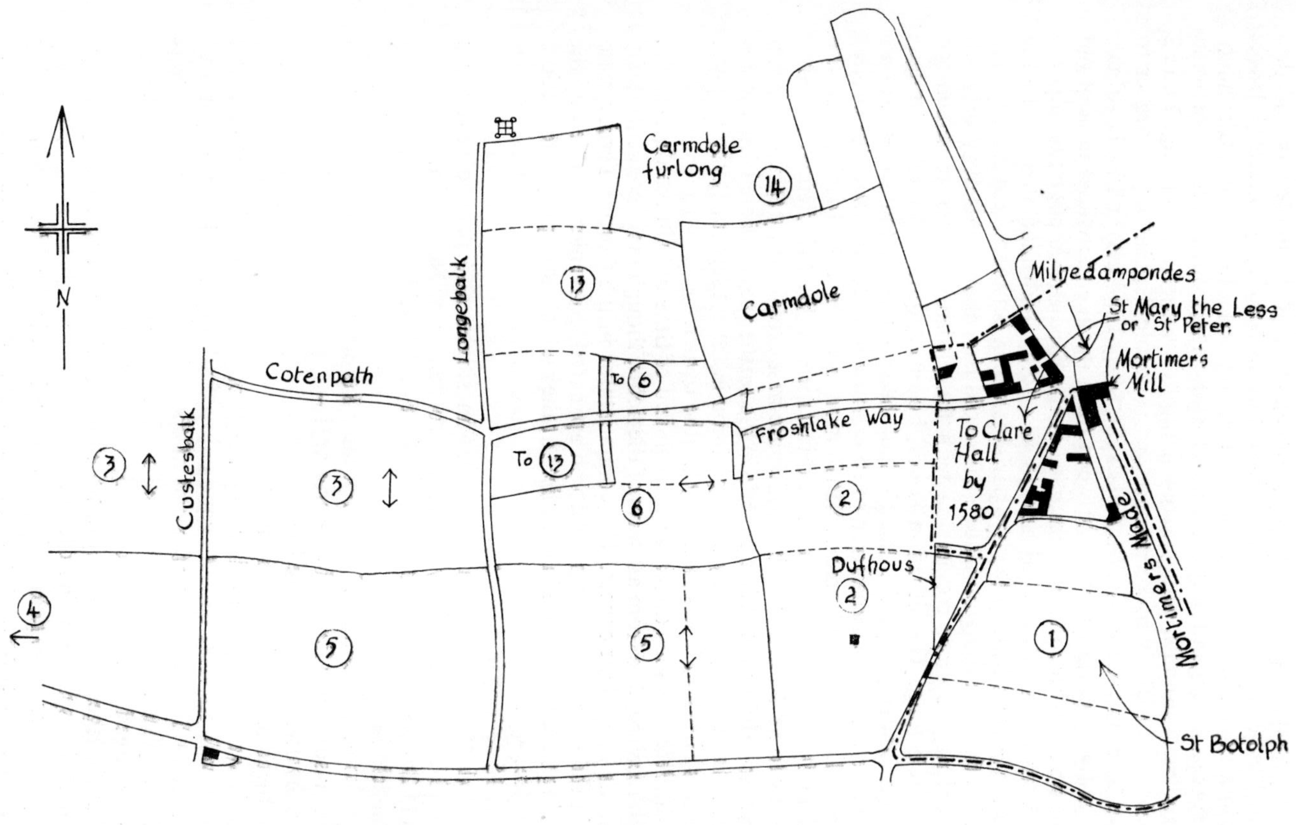

"Newnham Crofts" comprise the area of Qs 1, 2, part of 3, 5 and 6, up to Custesbulk

THE WEST FIELDS OF CAMBRIDGE

The distribution of tithe-ownership, like that of land-holding, shows a great difference in pattern between the Newnham/Carmefield and the rest of the West Field. The heading to the Carmefield sections of the Corpus Terrier explains that certain sections in Old Newnham Croft marked "Par" must be equally divided between Radegund and Peterhouse, and certain other sections between St. Giles and St. Botolph's, while the remainder of the selions in furlongs 1 and 2 of this Field tithe entirely to Peterhouse or St. Botolph's, the larger number to Peterhouse. Since the parishes of St. Peter's (St. Mary-the-Less) and St. Botolph's each have outlying portions which between them include the village area of old Newnham, we have here the only instance in the West Fields of something like parochial boundaries—that is to say, in the "old" Newnham field both the arable and the Houses tithe to the two parishes which between them cater for the area "outside the Trumpington Gate". These old Newnham fields obviously still have an ancient and close connection with the parish churches to which the village houses belong. The situation is not, however, quite as simple as might appear. The tithes of old St. Peter's had been appropriated to St. John's Hospital before they were transferred to Peterhouse in 1286, and were the subject of dispute and possibly of some redistribution between the Hospital and the College in 1339–40. There are one or two instances in the Middlefield of tithes going directly to St. John's Hospital which are best explained by the assumption that they are in origin tithes of St. Peter's in the further parts of the Fields which were adjudged to the Hospital by Bishop Simon in the settlement of the dispute. For St. John's Hospital is not normally a tithe owner.

Unlike St. Peter's, which appears as a recipient of tithe only in the Carmefield, St. Botolph receives tithe from all parts of the fields, though more concentratedly in the S.E., as one might expect. The commentator in the Corpus Terrier, who put *Nota* in the margin every time Botolph appeared, conveniently added up the selions paying tithe to Botolph in each field at the end of the appropriate section. His figures are:—

In Grithow, 15 selions in 4 pieces.
In Middlefield 21 selions, mainly scattered in singles and pairs except for a block of 10 in Q.22 owned by Nicholas Crocheman of Newnham.
In Littlefield 26 selions though it is the smallest of the fields, and
In Carmefield 28 selions.

If one reckons the selions of the area later known as Little Carmefield with the two other southerly fields, the figures are:—
To the Northwestern fields 26 selions
To the Southeastern fields 64 selions.

St. Botolph's was peculiar in other ways. Although the Church was anciently given to Barnwell, its vicar collected and enjoyed all the tithes in return for a £4 cash payment to the Priory. By the end of the Middle Ages this payment was extinguished and he had worked himself back into the position of Rector. The advowson belonged to Corpus from 1353, hence the great interest in Botolph tithes in that College. The church also tithed in the East Fields, the only church in the West Fields system which did so.

The Round Church also held an individual position in the West Fields tithing system, although its advowson, and presumably also its tithe, had belonged to

Barnwell from the early 13th century. The strips from which it drew its tithe are mainly in the central area of the old Portefield and in the Southern parts of the Fields. As has been mentioned, all the University lands tithe to *Rotund* and it is probable that Nigel de Thornton's benefaction represents the property of one of the original members of the Fraternity of the Holy Sepulchre which built the church (c. 1130). At this time, the Abbey of Ramsey, which granted the site, reserved its "ancient rights", but what these were is unspecified. If they included tithes, these had been transferred to Barnwell by purchase or exchange by c. 1200.

The history of the Round Church illustrates one of the difficulties we encounter when trying to discern early tithing systems. That pre-Conquest churches like St. Peter's and St. Botolph's should have 'signs of antiquity' (as Maitland calls them) about their tithing arrangements is not surprising. What he has to say about tithe distribution is perhaps relevant; "another sign of antiquity catches our eye. The distribution of the right to tithe is as intricately irrational as the distribution of proprietary rights. . . All is in wild disorder and seems plainly to tell of a time when men 'went with their land' to which church they pleased"; We may not go all the way with Maitland in the postulation of an early time of such chaotic freedom as is here described, but imagine it more likely that the earlier 'norm' was for a man to pay the tithes of all his holdings to the church of the parish in which his principal messuage lay. Yet the system seems still to have been flexible enough at the time of the foundation of the Round Church for townsmen to have been able to build a church which then acquired its own parish, and to endow it with tithe from their family lands.

When the history of St. Sepulchre's comes to be investigated as fully as that of St. Botolph's has been, it may turn out that the arrangements for the tithe payments of the two churches were similar in the thirteenth century, namely, the vicar collecting all tithes in return for a fixed cash payment to Barnwell. The entries in the Barnwell Book appear to be very similar. In the mid thirteenth century, the burgess William Tuillet founded a chantry of the Virgin Mary in the Church and the land he gave for its support is mostly found in Qs. 2–4 of Grithowfield. By this time, as shown by the Terrier, tithing of a selion had become firmly attached to that selion, regardless of who held it. Benefactions of land to religious houses or churches do not carry tithe with them unless a special arrangement is made with the tithe-owner. The only way an institution could now acquire tithe was by acquiring the church to which they were due and then appropriating them; hence the fierce competition for churches, especially in Cambridge when the Colleges came on the scene. Throughout the Terrier it will be noticed that the chantries and the colleges (charities fashionable in the late 13th and 14th centuries) and the local religious houses, and the Hospital (fashionable in the twelfth and early thirteenth centuries) do not receive tithe (except in so far as they have appropriate churches) but pay it to other religious bodies. The resulting little packets of receipts turn up in most college archives. Yet here and there in the Terrier we find traces of more ancient arrangements, for instance where "seynt gyles aker" or "seynt gylis rood", paying tithe to EGID, may represent an early donation of land that carried its own tithe.

THE GENESIS OF THE "BACKS"

In view of the number of plans showing conjectural reconstructions of the Town and its environs in the Middle Ages which continue to be reproduced and used as if they had documentary authority, it would be convenient to summarize here what can be learned from the Terrier, its successors and supporting documents, about the area immediately to the West of the River Cam. The earliest map of the Town to be reliably surveyed and drawn in the modern way is that of David Loggan of 1688. Before this we have maps showing details pictorially, of which the largest and most accurate is the Hammond map of 1592. But these maps all contain features which documentary evidence shows to have been relatively recent at the time they were drawn. Consequently they can be misleading if taken uncritically as the basis for a reconstruction of the medieval topography.

With regard to the main course of the River Cam, both maps show the river with rather more loops and bends than at present, and given the tendency of men to straighten and rationalize watercourses in the interests of drainage and navigation, it is reasonable to assume that the main loops and meanderings shown by Hammond were there in the Middle Ages. Moreover even the vestigial ones probably at one time carried more water but were robbed of their supply by artificial cuts and by the widening and deepening of the favoured channels. Two loops only need to be considered here as major topographical features. The first carried the water from the Newnham Mill directly across into what is now a mere ditch around the garden and new buildings of Queens' College. This was as large, or nearly as large, as the watercourse coming from the King' Mill under the College walls. The second loop was at Garrett Hostel Green, then an island, and it is possible that the left hand watercourse, now the main river, was artificially made in the interests of navigation. Between these two loops, the modern river has been carefully canalised and considerably banked up on the townward side. But barring a few minor deviations here and there its course was substantially as it is to-day.

This was not the case with the Bin Brook, the exact lower course of which has been a subject of debate. The modern course bears every sign of being an artificial cut, albeit an early one, if only because its route is improbable, bending back to cut into the 25 ft. contour, and its banks consist of a considerable amount of made-up ground. We believe the true original course to have been the winding stream, marked on old maps as "St. John's Ditch", and depicted in the Loggan print of that College dividing the fish-ponds from the meadow.[1]

Apart from geographical probability—course, direction and fall—there are sound documentary reasons for believing it to have been a very ancient boundary, which is not the case with the present stream.

1 D. Loggan *Cantabrigia Illustrata* ed. J. W. Clark, Cambridge 1905. reproduced also in Willis and Clark *Arch. Hist.,* Vol. II pp. 234–6.

The John's Ditch or Old Bin Brook (as we prefer to call it) was from early times to the nineteenth century a parish boundary. It is marked as such in the Baker map of 1830, and later maps of the nineteenth century continue to show a wavy boundary line through the middle of St. John's New Court. Moreover it is the boundary of St. Peter's "ultra pontem", a parish whose bounds became fossilised in early medieval times, so that they represent even more ancient limits. St. Peter's consists of three curiously detached portions, which make sense if they are regarded as between them containing all the inhabited areas at or near the southwest corner of the old walled Upper Town at the time that they were drawn. The lowest-lying portion was called the "Marsh" and its bounds were those, not only of the parish but also of the whole Upper Town. It is reasonable to suppose that the natural features chosen for such limits were notable ones.

The land to the northeast of the Old Bin Brook was not only all in St. Peter's parish on both sides of the present cut (or new Bin Brook, as it is convenient to call it), but it all belonged to one owner in the thirteenth century, namely Hervey Dunning. We believe that this land is all part of a single water-meadow specifically referred to in the Merton Charter dealing with the Stone House and its appurtenances by the phrase *cum hulmo*.[1] The meadow both north and south of the cut passed to Merton College, and the southerly portion was granted to St. John's Hospital by Henry VI in 1448. The piece of ground beyond the shoulder of the cut was sold by Eustace Dunning to his brother Thomas, by a charter recorded in the Cartulary of Sir John de Cambridge.[2]

Below this again, close to the foot of the bridge, was the home of Andrew the Fisher and his family, later called Fisher's Lane. His claims had to be bought off. In the quitclaim, also in the Cartulary, Andrew renounces all claim to the land "retro molendino" about to be transferred between the two Dunnings. Now the Dunnings had no known watermill, although later a horse-mill is mentioned. Is it possible that verbal usage, persistently conservative, gives us a clue to the reason why the new cut was made, even though there was no working mill there in the thirteenth century? If such a supposition is tenable, it is tantalising to speculate that here we have the ghost of the third mill of Domesday Cambridge,[3] that of Picot the Sheriff, which caused the townsmen's complaint.

Picot, Sheriff on Castle Hill, was likely to build a mill where it was convenient for him, and this site is the nearest point to Castle Hill at which one could make a mill cut. We have to remember that the Bin Brook probably carried more water than it does to-day, and its flow could have been reinforced above the bend by water from Castle Hill (perhaps in part channelled from the mysterious "Watercourse called Cambridge" about which this is not place to speculate further). We know that he removed pasture from the townsmen. Was this the water-meadow? It is odd that the water-meadow south of the Bin Brook should be common pasture, and north of the same should belong to one owner. We know that Picot was made to remove the millstones but there is no record that he restored the pasture. It does not seem unreasonable that it should eventually

1   M.1545. See p. 61 above.
2   C.C.C. Cartulary of Sir John de Cambridge fo. 33 et seq.
3   *T. & B.*, p. 190.

pass from Picot to whoever in practice succeeded him as the greatest man on Castle Hill. In 1200 this was Sir Baldwin Blangernon, and we know how the Stone House and its appurtenances passed from Blangernon to the Dunnings. But without necessarily accepting mere fascinating speculations, it is possible to maintain that the Old Bin Brook is the more ancient stream and a very ancient boundary.

In the Middle Ages, the area immediately adjacent to the west bank of the Cam, from just below Newnham Mill to the point where the Old Bin Brook entered the river, was known as Long Green. It was continuous open pasture for its entire length, unplanted with trees and unbroken by any formal watercourses. In appearance it must have been similar to Grantchester Meadows to-day, and like Grantchester Meadows, only provided a continuous walk in dry weather, for it would have carried the run-off from the Carme Field at various points and have been liable to flooding, since it all lay below the 25 ft. contour. It immediately joined the arable on its western edge, no road or watercourse intervening. At the Newnham end, however, the arable was separated from the common way and the common green by the "pratum Mortimer", Mortimer's Meadow, recorded by the earliest extant charter as meadow or pasture belonging to the Mortimer manor. It is probable that this meadow was originally separated from the common by hurdles or ditches; by 1592 it had a wall, and later it was planted with hedges and was called Butcher's Close. Even on Hammond's map it is shown as superior to the common pasture of the Long Green, for while the latter carried scrawny cattle, the Mortimer pasture is filled with superb prancing horses. Hammond's map also shows clearly that there was no through road marking the edge of the Long Green even by 1592; where Newnham High Street forks right to cross the river by the Small Bridges there is just a drove on to the Green.

Although the most westerly portions of the Hammond map tend to disappear under labels at crucial points, the uneven outline of the cornfields depicted on it suggest that there has been some cutting back or lapsing of the arable along parts of the eastern edge of the Carmefield. A detailed comparison between the Corpus Terrier and later maps shows that this has indeed been the case, particularly towards the northern end and that Maitland was correct in his belief that the Johnian Wilderness had once formed part of the West Fields arable.[1] In particular, the easterly selions of furlongs 10 and 12 had been eroded; by how much the sixteen selions of the Clerks of Merton in furlong 12 had been reduced it is hard to judge, but the six selions *sine balca* of furlong 10 marked in the margin of the Terrier as "regal coll." have vanished by 1789. Presumably the tranverse selions of furlong 11 have likewise been cut off at their eastern heads, and it would appear from Hammond's map that only the first two owners, the Hospital and Corpus Christi College, have maintained arable cultivation. The last two owners, King's College (replacing the Prior of Huntingdon) and Corpus again, have converted their arable to leys indistinguishable by any fencing from the common pasture. Similarly, across the Bin Brook, about half of the ninth furlong of Little Field has been taken out of arable cultivation. The original reason for this abandonment may have been the wetness of the ground,

1 T. & B., p. 122.

The Carmefield where it abutted on the Long Green (as in MS.2601)

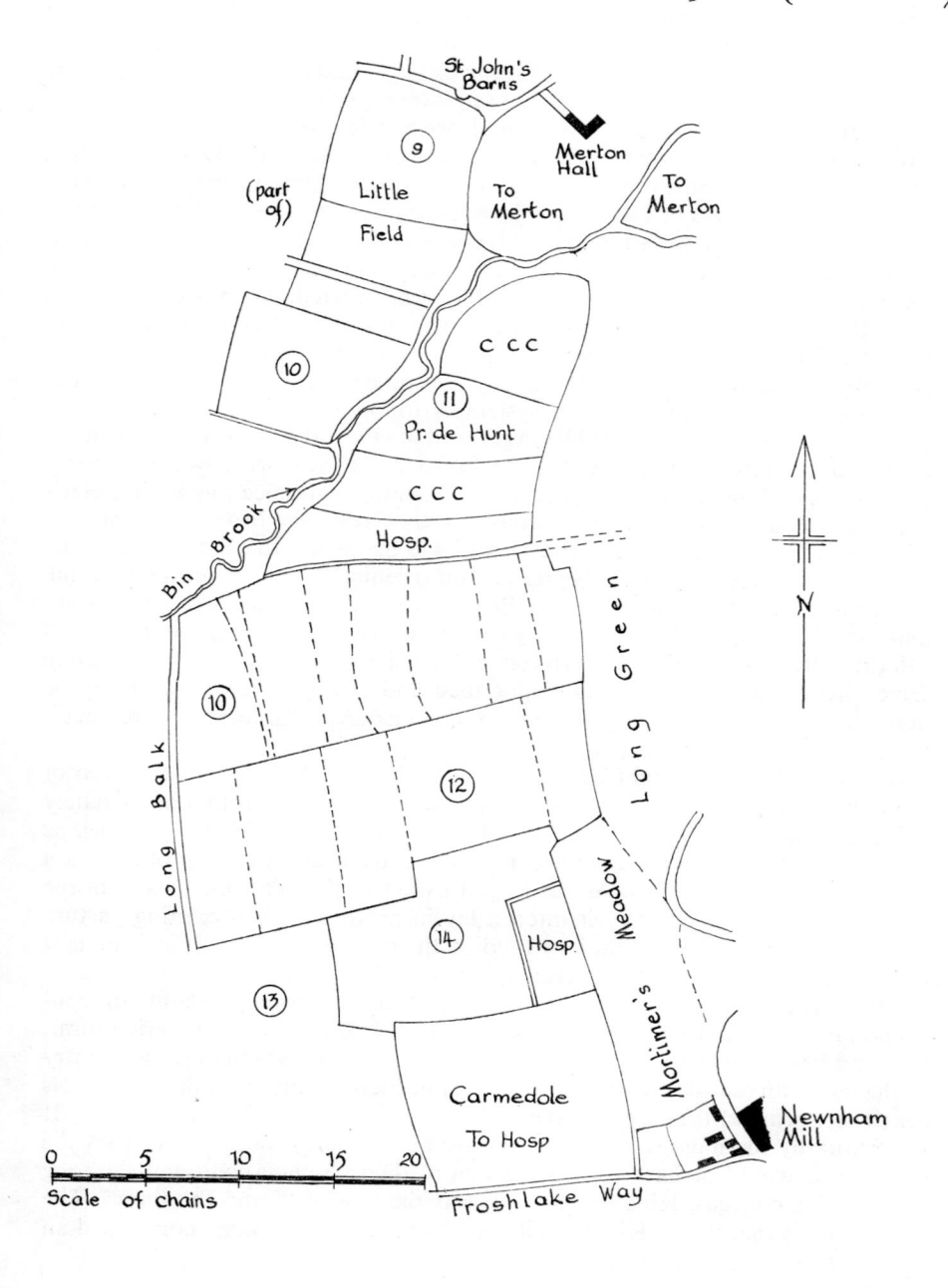

for all the area close to the Bin Brook is low-lying. That it remained out of cultivation was probably due to two factors which did not become important until the sixteenth century, namely the encroachment of the Colleges upon the common, and the increasing use of wheeled vehicles.

There is no convincing evidence that the Colleges had in any way "jumped the river" to appropriate portions of the Long Green before the sixteenth century. Willis and Clark, indeed, infer from the Letters Patent of Henry VI of 1448, by which the Hospital obtained the Fishponds area north of the Old Bin Brook, that the piece later called St. John's Meadow, on the Long Green, was already a garden in the hands of the Hospital.[1] But they were not aware, and could not be, of the peculiar orientation generally used not only to describe arable lands in the West Fields, but also for urban properties in the northern and western sections of the town. This system, it will be remembered, equates the direction of both Huntingdon and Madingley roads with north. (Indeed almost any point over the river is north, and conversely any point towards the centre of the town is south; one's neighbours to left and right become east and west. All properties are treated as rectangular). This system is found in charters of Cambridge properties throughout the Middle Ages and was clearly in current use in St. John's and Merton. Consequently, the Fishponds area (formerly the property of Merton, and more anciently part of the Dunning watermeadow in St. Peter's parish), which lay in the angle formed by the New Bin Brook and the Old Bin Brook, is described in the charter of 1448 as having the Bin Brook on its "north" and "west" sides. Applying the old orientation this would refer to the two arms of the Old and New Bin Brook, while the gardens on the "south" and "east" would be the Hospital gardens known to exist in what we would call the northeast corner by the riverside (later site of Nutter's Yard). The Long Green lies outside all the bounds described and in any case would hardly be described as a "garden", for the gardens of medieval Cambridge were small, walled and intensely cultivated.

Another point that might be made is that while the Master and Brethren of the Hospital had an obvious interest in acquiring a piece of ground immediately adjacent to the Hospital buildings for use as fish-ponds and herb gardens, as landowners they had absolutely nothing to gain from the enclosure of common pasture of the Long Green. As the largest owner in the West Fields and in the Carme Field in particular their interest lay in protecting their existing pasture rights. Later we find the Johnians loud in their protests when their dominant position is threatened by the encroachment of other colleges.

The motives behind the acquisition of portions of common land for conversion into private tree-planted walks were clearly social and not agricultural. We have here another aspect of the "emparking" fashion among the new gentry of the sixteenth century. As the colleges became increasingly the finishing schools for the offspring of these gentry, they felt the desirability of similar amenities. It was naturally the most socially competitive and wealthy colleges which set the example. It would be hard to prove which of the two main contestants among the riverside colleges, King's or John's, was the first to "jump the Cam". The probability is that it was King's College, which, legally speaking, committed an

1 *Arch. Hist.* Vol. II. pp. 235–6.

unprecedented act by purchasing a substantial portion of Long Green from the Corporation of the Town. The St. John's Terrier of 1566 describes the position in terms which are confused and probably disapproving:
> Long Meadow is without Kinges College and sometime all the gardens and closes that the Kings College hath now without their great bridge toward Cotton Way called Long Meadowe sometime before the Kinges College had it purchased of the commonalty of Cambridge.[1]

To a Johnian, the head and front of King's offending was not just the conversion of part of the former common pasture into a private walk, enclosed by ditches, planted with trees and linked to the main college by a new bridge over the river. St. John's, once it had become a college, showed no inclination to be left behind in any of these amenities, and indeed made good speed to provide them for itself, by the simple process of running a watercourse round the end of the Long Green opposite its own main buildings. John's may even have preceded King's in the enterprise. What upset the Johnians was the tacit admission, implied in the purchase from the Town, that the Town officials had the right to dispose of the common pasture belonging traditionally to the owners of the arable of the West Fields, and in particular of the Carme Field, now being called the College Field by St. John's. Johnians would not stoop to ask or buy permission from the Town, whose rights in the Green they utterly contested. Hence the absence of written record as to when the John's Meadow slipped into sole possession of the College.

This reconstruction of events is inevitably in part conjecture; what is undoubted fact is that on the early maps of sixteenth century Cambridge, Lyne's of 1574, Braun's (from the same basic design) of 1590, and in most detail Hammond's of 1592, the only two colleges who have as yet thrown bridges over the river to connect with areas surrounded by watercourses and planted with trees, are King's and John's. The former, which boasts a little water-surrounded building[2] as well as an elaborate plantation, is labelled, "Kynges College Back Sides", and the latter, a simple central avenue, is "Jhon's Walkes".

It is nowhere explicitly stated that King's College had to provide any sort of compensation for the loss of pasture by throwing some of its arable into leys. But the fact that King's College land in the adjacent arable goes out of cultivation at about this time suggests that this was the case. It would also appear, from such mapping as is possible, that King's took the existing Green to its outer limit when making their enclosing watercourse. (The present gate stands right on the road). This left two shoulders of arable land (furlongs 11 and 12), sticking out between the point of entry for cattle and the new green leys. Soon these were abandoned, probably because they were too awkward and expensive to protect. This cannot have improved the already strained relations between King's and the owners of the 'lost' pieces, Merton, Corpus and John's.

The portion of the King's College Backsides immediately behind Clare was subsequently made over to that college by King's, but this did not affect the amount of Long Green available for common pasture.

1 See Appendix A.
2 Possibly an icehouse. It was certainly not a medieval moated house.

The conflicting forces within St. John's, its interests as a new college as against those as a West Fields owner, go far to explain the fierceness and the apparent inconsistency (as it seemed to Maitland) of the Johnian attitude, when in 1610 Trinity College showed signs of joining the "Backs" league. On the one hand St. John's was negotiating with Corpus Christi College for a long lease of that portion of its former arable close to the Bin Brook which had lapsed into rough pasture and gravel diggings. St. John's wanted the land for a Fellows' Garden and a new bowling green, even though it was a fair distance from the college, and we may note that readiness to acquire land from others for ponds or gardens was accompanied by reluctance to diminish the pasture of their own farming tenants. On the other hand, St. John's as West Fields owner was highly affronted by Trinity's method of purchase from the Town, and made common cause with the protesters who insisted that those deprived of pasture rights should be compensated elsewhere. The result, in 1613, was the grant by Trinity of Parker's Piece as a substitute open space. This would seem rather a poor substitute as it was obviously less lush grazing than the riverside pasture, and of little use to owners whose interests lay in the West Fields. But at least St. John's had never had to compensate for its own enclosure, and in 1622 went so far as to swallow its pride and make its own bargain with the Town for the last little bit of rough ground between the ex-Corpus land and the Bin Brook.[1]

Trinity College, like King's before, put its new encompassing ditches at the furthest limit of the Long Green, leaving only a narrow access way from the John's meadow at the north end, and extending further over the south end than King's had previously done. This was possibly because furlong eleven was ploughed transversely and Trinity was able to take in the former headlands.

The lease by which St. John's took over the former Corpus arable mentions that the land extends from Long Green, "in parte inclosed and now belonging to Trinitie Colledg ...over the Common waie".[2] From this we learn that a generally recognised route has by now been established from Garret Hostel Lane over the northern half of the Green, as well as the earlier route over the southern half (shown on the sixteenth century maps). This way crossed the Bin Brook by a ford, clearly shown by Hammond as a widening of the brook at the point where it was later (1822) crossed by a bridge. It would seem that the making of such a ford only become possible as arable was cut back to give access from both sides on the firmer, more gravelly ground, and the spoil of the gravel diggings was probably used to make the ford shallow. Once established, the new route had obvious advantages over the old Barton Way, which seems to have declined from about this time, as the new route was increasingly used. The new, and not the old route was eventually selected for the eighteenth century turnpike. (It is fascinating to speculate on the subsequent traffic flow and patterns of suburban development in West Cambridge had the old Roman road been chosen).

It is surely not coincidence that in 1626, two years after the last major enclosures by Trinity and St. John's, Town and University joined forces to stint the

1   A detailed description of the conversion of the Corpus lease to purchase, further acquisitions, and the making of the new garden, together with incidental alterations to the course of the Bin Brook, is given in *The Eagle,* MCMLI, pp. 300–04.
2   *Arch. Hist.* II. p. 238.

use of common more severely than ever before. Continual and increasing passage over the narrow strip of compensatory green must have further reduced its usefulness for grazing. The turnpike must finally have given the *coup de grace* to the practicable use of this green for pasture. In the 1770's, about the time of the making of the turnpike, the last of the compensatory green was planted with an elm avenue under the auspices of Trinity College. The Fellow's horse had replaced the commoner's cow. Alone of the riverside colleges, Queens', with an island for its gardens and walks, never took the necessary steps for the complete appropriation of the adjacent common green. Only in the far southerly corner of the old Long Green, the part known as Queen's Green retains approximately its former appearance and use. Elsewhere, the Backs as we know them are entirely the creation of the three largest colleges, and, equally with their buildings, have effaced the works of man and nature which preceded them.

THE TEXT

Grythowefeld

1

[fol. 1r.]
♭

Memorandum quod garbe decimalium de omnibus selionibus iacentibus in Grithowfeld qui dicuntur. Par. ut infrascriptum est debent eque partiri inter Radegundam et Egidium videlicet .j. garbam Radegunde et.j.garbam Egidii exceptis garbis decimalium de.iij.selionibus hospitalis sancti Johanis evangeliste Cantabrigie que debent eque partiri inter Egidium Botulphum et Rotundam ut patet infra

♭   1   Quarentena abuttans ad capud suum australe contra le Cleypittes ad Asshwykston *at the castell end*
Seliones illius quarentene debent computari ad capud eorum boriale
viij.seliones tamen aliqui eorum sunt gores cum.j.longa butta iacentes ex parte occidentale ad capud eorum australe de terra Hospitalis sancti Johanis evangeliste Cantabrigie et sunt primi et orientalissimi illius quarentene et iuxta huntyngdonweye   *vj acr'* Egid.

a   iiij.seliones de terra Mortimeres iuxta/quorum.ij.primi seliones sunt curtiores aliis selionibus antedictis ac etiam aliis selionibus sequentibus ad capud eorum australe et sunt ultimi et occidentalissimi illius quarentene et iuxta Grythowwey   *ij acr'* Rad.

a   iiij.seliones de terra Mortimerez iacentes transversi ad capud boriale quarentene ultimo dicte et extendunt se in longitudine a Huntyngdonwei predicta usque ad Grythowwey predictam et sunt una quarentena per se   *ij acr'* Rad.

+
♭   2   Quarentena abuttans ad capud suum australe super.iiij^{or}. seliones de Mortimeres ultimo dictos

+   *Ashwykston ys by the hye crosse at the castell end in huntyingdon weye on the weste sowth part of the seyd cross and ytt ys lyke a stompyd cross*
ix seliones et una butta vocatur sal iuxta castrum et ij forera tota crofta de terris de merton hall ex parte boriali et potius occidentale castri et est una quarentena per se ij acr'

1a

[fol. 1v.]

seliones illius quarentene debent computari ad capud eorum australe (*deleted*)/boriale

a   vj.seliones Prioris de Huntyngdon tamen aliqui eorum sunt gores primi et orientalissimi illius quarentene et iuxta Huntyngdonwey predictam   *j acr di* Rad.

GRYTHOWEFELD

Inter. ij. balcas *cχ*	⎧ j.selio Thome Bolle quondom Johanis de Toft iuxta *cantarie beate marie j rod* Rad. ⎨ j.selio collegii Corporis Christi quondam Thome de Cante- ⎪ brigia iuxta	*b di acr'* Rad.
	⎩ j.selio Thome Boll quondam Johanis de Toft iuxta *cantarie marie j rod* Rad.	
mo	x.seliones Stephani Morys senioris quondam Johanis de Berton iuxta	*iij acr' et di* Egid.
	⎧ j.selio Thome Boll quondam Johanis de Toft iuxta et est longior ⎪ aliis selionibus antedictis et in parte forera ad capud suum boriale	*di acr'* Rad.
sine balca	⎨ v.seliones hospitalis predicti iuxta	*j acr' j rod* Egid.
cχ c	⎪ vj.seliones Collegii predicti quondam Galfridi Seman iuxta et sunt ⎪ ultimi et occidentalissimi illius quarentene et iuxta Grythowweye	
	⎩ predictam *ij acre*	*b j acr' di* Egid.
a	x.seliones cum sua forera ad capud eorum australe de terra hospitalis predicti iacentes inter Grythowweye predictam ex una parte et le Stoupendcrouch Wey predictam ex altera parte et abuttant ad capud eorum australe super le Cleypittes antedictum ex oposito Horhill et et sunt una quarentena per se	*vj acr'* Egid.
⟨ 3 *a)*	Quarentena iacens transversa ad capud Boriale.x.selionum hospitalis predicti ultimo dictorum // Seliones illius quarentene debent computari ad capud eorum orientale	
	ij.seliones hospitalis predicti quorum primus selio est forera *ad decem ultimos* et sunt primi et australissimi illius quarentene et abuttant ad capud eorum orientale super Grythowweye predictam et ad capud eorum occidentale super	

[fol. 2r.]   2

	le Stoupendcrouchweye predictam	*iij rod* Rad.
	v.seliones Thome Morys iuxta	*lovell of chesterton ij acr'* Egid.
	iij.seliones Stephani Morys senioris quondam patris sui qui vocatur le Sponyaker iuxta	*j acr'* Rad.
	ij.seliones Thome Morys iuxta	*j acr'* Egid.
	iij.seliones hospitalis predicti iuxta	*j acr' j rod* Egid.
	ij.seliones Stephani Morys senioris quondam Johanis de Berton iuxta	*j acr' j rod* Egid.
cχ c	ij.seliones Collegii predicti quondam Thome de Cantebrigia	*b iij rod* Rad.
	⎧ iij.seliones hospitalis predicti iuxta .Par. videlicet	*j acr'*
	⎪ Egidii.j.garbam Botulphi.j.garbam et Rotunde.j.garbam	Botulph' *nota*
.b.	⎨ j.selio monialum de Beche iuxta	*di acr'* Par.
Inter.ij. balcas	⎪ j.selio hospitalis predicti iuxta	*iij rodes sive di acr'* Egid.
cχc	⎩ j.selio Collegii predicti quondam Thome de Cantebrigia iuxta	*b di acr'* Rad.
	vij.seliones Stephani Morys senioris quondam Johanis de berton iuxta	*iij acr' di* Egid.
	ij.seliones dicti Stephani quondam Galfridi Seman qui vocatur le Blakaker iuxta	*ij acr'* Par.
	ij.seliones dicti Stephani quondam dicti Galfridi iuxta quorum ultimus selio est in parte forera ad capud suum occidentale	*ij acr'* Egid.
	ij.seliones dicti Stephani quondam Johanis Redhod iuxta et sunt curtiores aliis selionibus antedictis ad capud eorum occidentale et abuttant super foreram Collegii predicti quondam Galfridi Seman ad capud predictum	*v rod* Par.

89

THE WEST FIELDS OF CAMBRIDGE

sine balca
        ij.seliones hospitalis predicti iuxta et abuttant super foreram
        predictam   *di acr'* Egid.
        iiij.seliones Thome Morys iuxta et abuttant super foreram predictam
        et sunt ultimi et borialissimi illius quarentene quorum ultimus selio
        est forera   *ij acr'* Egid.

♭   4   Quarentena abuttans ad capud suum orientale super   *iuxta le cawseye*
        Huntyngdonweye // Seliones illius quarentene debent computari ad
        capud eorum orientale

2a

[fol. 2v.]

cant' marie
[sic] *ad fori*
sine balca
        ij.seliones hospitalis predicti primi et australissimi illius quarentene
        quorum primus selio est forera   *j acr'* Rad.
        j.selio Thome Bolle quondam Johanis de Toft iuxta   *di acr'* Rad.
        j.selio hospitalis predicti iuxta   *di acr'* Egid.
        ij.seliones cantarie beate marie ecclesie sancti sepulcri iuxta et ambo
        sunt forere in parte ad capud eorum occidentale et sunt longiores
        aliis selionibus antedictis ac etiam aliis selionibus sequentibus ad capud
        predictum et extendunt se in longitudine a Huntyngdonweye predicta
        usque litilmer   *ij acr'* Egid.
        viij.seliones Roberti de Brygham iuxta   *ij acr'* Egid.

woods

Inter.ij. balcas
   –o
        ij.seliones ocupati per dictum Robertum de terra quondam Johanis
        Berton prout patet per diversos libros de decimis de Bernewell iuxta   *iij rod* Egid.
        ij.seliones ocupati per Stephanum Morys seniorem de terra capelle
        de Steresbregg prout patet per libros predictos iuxta   *j acr'* Par.

    mo
        iiij.seliones qui solebant esse.iiij.dicti Stephani quondam patris sui
        iuxta   *j acr'* Egid.

croft
        j.selio Thome Bolle quondam Johanis de Toft iuxta   *di acr'* Rad.
        iij.seliones de terra capelle de Steresbreg iuxta   *j acr' et di* Par.
        iij.seliones hospitalis predicti iuxta   *j acr' et di* Egid.

o
        j.selio qui solebat esse.ij.Thome atte Chirch de howes quondam
        Willelmi de Lavenham iuxta   *nunc hospitalis*   *di acr'* Rad.
        ij.seliones hospitalis predicti iuxta   *j acr'* Egid.
        j.selio qui solebat esse.ij.de terra capelle de Steresbreg iuxta   *j acr'* Par.
        j.selio Rogeri Harleston quondam Willelmi de Lolleworth iuxta   *di acr'* Par.
        iiij.seliones hospitalis predicti iuxta   *ij acr' di* Par.
        iij.seliones eiusdem hospitalis iuxta   *j acr' di* Egid.

cχ   ↄ
        j.selio collegii predicti quondam Galfridi Seman iuxta   *b di acr'* Egid.
        ij.seliones sancti Egidii qui vocatur Sengilakir iuxta   *j acr'* Par.
        j.selio Cantarie beate Marie ecclesie sancti sepulcri iuxta   *di acr'* Egid.
        j.selio Thome de Marbilthorp iuxta   *clar' hall*   *di acr'* Par.
        j.selio Cantarie beate Marie ecclesie sancti sepulcri iuxta   *di acr'* Egid.

cχ   ↄ
        j.selio Collegii predicti quondam Thome de Cantabrigia iuxta   *b di acr'* Rad.

3

[fol. 3r.]

        j.selio hospitalis predicti iuxta   *di acr'* Egid.

cχc
        j.selio Collegii predicti quondam Galfridi Seman iuxta   *di acr' b* Egid.
        ij.seliones Rogeri de Harleston quondam Willelmi Ward iuxta   *j acr'* Egid.

    mo
        ij.seliones Stephani Morys junioris quondam Johanis Pittok
        iuxta   *j acr'* Par.

GRYTHOWEFELD

	j.selio de terra capelle de Steresbreg iuxta	di acr' Par.
	j.selio Cantarie beate Marie ecclesie sancti sepulcri iuxta	di acr' Egid.
mo	iij.seliones Thome Morys iuxta	j acr' di Egid.
	j.selio qui solebat esse.ij.Thome atte Chirch de Howes quondam Willelmi de Lavenham iuxta *hospitalis*	di acr' Rad.
	j.selio Rogeri de Harleston quondam Ricardi Tableter iuxta	di acr' Rad.
	ij.seliones Thome de Marbilthorp iuxta *clar' hall*	v rod. Par.
	ij.seliones qui solebant esse.iiij^{or}. hospitalis predicti iuxta	j acr' Egid.
et b cχc	iiij.seliones qui solebant esse.viij.Collegii predicti quondam Thome de Cantebrigia iuxta	ij acr' b Rad.
	v.seliones Cantarie beate Marie ecclesie sancti sepulcri iuxta	ij acr' di Egid.
.b.	⎧ j.selio monialum de beche iuxta	ij rod.' Par.
Inter.ij. balcas	⎪ iij.seliones Cantarie beate Marie ecclesie sancti sepulcri iuxta	j acr' di Egid.
cχc	⎨ j.selio qui solebat esse.ij.collegii predicti quondam G. Seman	
Inter.ij. balcas	⎪ iuxta	iij rod b Rad.
mo	⎩ iiij.seliones Thome Morys iuxta	j acr' di Egid.
	ij.seliones quondam Roberti Seman iuxta *clar'hall*	j acr' Rad.
	iij.seliones hospitalis predicti iuxta	j acr' di Egid.
al ca	ij.seliones alborum canonicorum iuxta *al ca*	j acr' Par.
cχc	ij.seliones Collegii predicti quondam Galfridi Seman iuxta et sunt longiores aliis selionibus antedictis in duplum	ij acr' b Egid.
	iiij.seliones qui solebant esse.vj.Thome atte cherch de Howes quondam Willelmi de lavenham iuxta.quorum.ij.primi seliones abuttant super huntyngdonweye/et.ij.ultimi seliones sunt curtiores aliis selionibus antedictis ad capud eorum orientale et abuttant super mesuagium dicti Thome in le Howes quondam dicti Willelmi	ij acr' Rad.

3a

[fol. 3v.]

	j.selio Thome Jekke quondam patris sui iuxta	ij rod. Rad.
	j.selio dicti Thome quondam patris sui iuxta	ij rod Egid.
	iiij.seliones Henrici blancpayn de Gyrton quondam Johanis atte grene iuxta	ij acr' Rad.
	iij.seliones Thome Jekke iuxta quorum primus selio quondam Roberti Rye et.ij.ultimi seliones quondam Galfridi Seman de merton lond et iacent iuxta balcam que divisit campum de Cantabrigie a campo de Gyrton *et omnes isti seliones a duobus selionibus collegii corporis*	
::	*christi iam sunt in diversis clausuris infra le Howes et vix possunt distingui*	ij acr' Egid.
⌂ 5	Quarentena iacens transversa ad campum de Gyrton abuttans ad capud suum occidentale super melneweye / / Seliones illius quarentene debent computari ad capud eorum occidentale	
mo	viij.seliones Thome Morys primi et borialissimi illius quarentene quorum primus selio est quasi forera ad campum de Cotez et abuttat ad unum capud super foreram Thome atte chirch de howes et ad aliud capud super le Howescroft	iij acr' Egid.
	iij.seliones Prioris de huntyngdon iuxta et abuttant ad unum capud super croftam predictam et ad aliud capud super Melneweye	j acr' Rad.

THE WEST FIELDS OF CAMBRIDGE

mo	j.selio Thome Morys iuxta et abuttat super viam predictam	*j rod di* Egid.
§ cχc	j.selio Collegii predicti quondam Thome de Cantebrigia iuxta et	
Inter.ij. balcas	abuttat super viam predictam	*j rod di b* Rad.
clausa iam	j. selio Thome Jekke quondam Johanis blancpayn iuxta abuttans	
ecclesie beate	super viam dictam	*di acr'* Rad.
marie in foro		
	iiij.seliones Thome Morys iuxta et abuttant super viam predictam	*j acr' di* Egid.
Inter.ij. balcas	j.selio Rogeri de harleston quondam Willemi Ward abuttans super dictam viam	*di acr'* Egid.
	j.selio hospitalis predicti iuxta et abuttat super viam predictam	*di acr'* Rad.
mer'	v.seliones qui solebant esse.viij.de terra clericorum de merton iuxta et abuttant super viam predictam	*iiij acr'* Egid.
	iij.seliones Prioris de Huntyngdon iuxta et abuttant super viam predictam	*ij acr'* Rad.

[fol. 4r.]

4

mo	j.selio Thome morys iuxta et abuttat super viam predictam	*di acr'* Egid.
nunc j cχ c	ij.seliones Collegii predicti quondam Thome de Cantebrigia iuxta et abuttant super viam dictam quorum ultimus selio est curtior aliis selionibus antedictis ad capud suum orientale	*b di acr'* Rad.
	j.selio Thome atte chirch de Howes quondam Willelmi de lavenham iuxta et abuttat super viam predictam *galfridi taverner*	*di acr'* Rad.
mer'	viij.seliones qui solebant esse.x.de terra clericorum de merton iuxta et abuttant super viam predictam	*vj acr'* Egid.
	ij.seliones Rogeri de harliston quondam Ricardi Tableter iuxta et abuttant super dictam viam	*j acr'* Egid.
mo	ij.seliones Stephani Morys senioris quondam patris sui iuxta abuttantes	
Inter.ij. balcas	super predictam viam	*j acr'* Egid.
	ij.seliones Prioris de Bernewell iuxta et abuttant super predictam viam	*j acr'* Egid.
	j.selio cantarie beate marie ecclesie sepulcri iuxta et abuttat super viam predictam	*di acr'* Egid.
mer'	iij.seliones de terra clericorum de merton iuxta abuttantes super predictam viam	*j acr' di* Egid.
cantarie marie in foro	j.selio Roberti Longe iuxta in manu Prioris de bernewell et abuttat super viam predictam	*di acr'* Par.
mo	ij.seliones Thome Moris iuxta et abuttat super viam predictam	*j acr'* Egid.
mo	vj.seliones Stephani Morys senioris quondam patris sui iuxta et abuttant super dictam viam	*iij acr'* Egid.
	ij.seliones qui solebant esse.iiij.Thome atte cherch de Howes quondam Willelmi de lavenham iuxta et abuttant super viam predictam *galfridi taverner*	*ij acr'* Egid.
	iiij.seliones ocupat[i] per Priorem de bernewell iuxta et abuttant super predictam viam	*ij acr'* Rad.
	j.selio hospitalis predicti iuxta et abuttat super viam predictam	*di acr'* Rad.
Inter.ij. balc'	j.selio Thome atte cherch de Howes quondam Willelmi de lavenham iuxta et abuttat super viam predictam *ecclesie marie in foro*	*di acr'* Rad.
mer'	v.seliones qui solebant esse.vij.de terra clericorum de merton iuxta et abuttant super viam predictam	*iij acr' di* Egid.

GRYTHOWEFELD

Inter.ij.	j.selio hospitalis predicti iuxta et abuttat super predictam viam	di acr' Rad.
	j.selio Roberti Longe iuxta in manu Prioris de Bernewell et abuttat super viam predictam le greneplat mediante *gravell pytt'*	j rod Par.
	j.selio monialum sancte Radegunde iuxta et abuttat super viam predictam le greneplat predicto mediante *gravell pyttes*	j rod di Par.

[4a]

[fol. 4v.]

balcas	ij.seliones qui solebant esse.iij.Prioris de Bernewell iuxta et abuttant super viam predictam le greneplat predicto mediante quorum ultimus selio est in parte forera ad capud suum orientale	j acr' Egid.
mo Grythowe	ij.seliones Thome Morys extendentes se in longitudine ultra Grythowweye iuxta et sunt longiores aliis selionibus antedictis ad capud eorum occidentale et curtiores aliis selionibus antedictis ad capud eorum orientale et abuttant super viam predictam	j acr' Rad.
‖‖‖ mer' Inter.ij. balcas	j.selio de terra clericorum de merton iuxta et abuttat super Grythowepath directe ex oposito Grythowhill	j rod di Egid.
	j.selio Rogeri de harleston quondam Willelmi de Lolleworth iuxta et abuttat super Grythowepath	j rod di Par.
et predicti .vj. seliones iacent inter.ij. balcas woodes	iiij.seliones hospitalis predicti iuxta et abuttat super Grythowepath	ij acr' Par.
	ij.seliones Roberti Longe iuxta et abuttat super Grythowepath	j acr' Par.
	j.selio hospitalis predicti iuxta et abuttat super Grythowepath	j acr' Egid.
Inter.ij. balc'	j.selio Ricardi Tuliet iuxta et abuttat super Grythowepath	j acr' Rad.
	j.selio Prioris de huntyngdon iuxta et abuttat super Grythowepath	di acr' Rad.
mo	j.selio Stephani morys senioris quondam Rogeri yslep' iuxta et abuttat super Grythowepath	di acr' Rad.
mo	ij.seliones qui solebant esse.iij.T.Morys iuxta et abuttant super Grythowepath	ij acr' Egid.
.b. Inter.ij. balcas cχc	j.selio monialum de bech iuxta et abuttat super Grythowepath	di acr' Par.
	j.selio Collegii predicti quondam Thome de Cantebrigia iuxta abuttans super Grythowepath	j rod b Rad.
	ij.seliones qui solebant esse.iij.Rogeri de Harliston quondam Ricardi Tableter iuxta et abuttat super Grithowepath	ij acr' Egid.
cχc Inter.ij.balc'	j.selio collegii predicti quondam Thome de Cantebrigia iuxta abuttans super Grithowepath	j rod b Rad.
¶ .b.	j.selio monialum de bech iuxta et abuttat super Grithowepath	j rod Par.
mo	ij.seliones Stephani morys junioris quondam Galfridi Seman iuxta quorum primus selio abuttat super Grithowepath et secundus selio abuttat super viam Sancti neoti sicut iacet in the turn *iuxta hunelles closses*	ij acr' Par.
	iiij.seliones Roberti Tuliet iuxta iuxta (sic) et abuttant super viam predictam et sunt	

¶ *jacet destride ad viam sancti neoti*

5

[fol. 5r.]

	curtiores aliis selionibus antedictis ad capud eorum orientale	ij. acr' Rad.
	j.selio Johanis de Weston quondam hugonis Pittok iuxta et abuttat super viam predictam *clar' hall*	j acr' Egid.

93

THE WEST FIELDS OF CAMBRIDGE

Inter.ij. balcas *kateryne hall*	⎧ iiij.seliones Rogeri de Harliston quondam Willelmi de Lolleworth ⎪ iuxta et abuttant super viam predictam ⎨ j.selio Prioris de Huntyngdon iuxta et abuttat super viam predictam ⎪ j.selio Thome atte cherch de Howes quondam Willelmi lavenham ⎩ iuxta	*ij acr'* Par. *j rod'* Rad. *j rod'* Rad.
Inter.ij. balcas	iij.seliones Johanis de Weston quondam Hugonis Pittok iuxta et	
al ca	⎧ abuttant super viam predictam *clar' hall* ⎨ iij.seliones alborum canonicorum iuxta et abuttant super viam ⎩ predictam	*ij acr'* Egid. *j acr'* Par.
mo	v.seliones Stephani Morys senioris quondam Willelmi de Bekeswell.... iuxta et abuttant super viam predictam iiij.seliones quondam Willelmi de lavenham iuxta abuttantes super viam predictam	 *ij acr' di* Egid. *ij acr' di* Egid.
nota cχc	ij.seliones Collegii predicti quondam Thome de Cantebrigia abuttantes super dictam viam iij.seliones Johanis de Weston quondam Hugonis Pyttok iuxta et abuttant super viam predictam quorum ultimus selio est forera et sunt ultimi et australissimi illius quarentene et iuxta le stoupendecrouch *clar' hall*	 *iij rod b* Botul. *j acr' di* Egid.
♭ 6	Quarentena vocata le Stoupendecrouchfurlong Seliones illius quarentene debent computari ad capud eorum australe *ista quarentena jacet* *iuxta viam sancti neoti et prius iuxta quarentenam ultimo dictam.*	
mer' *stoupyng* ✠	vj.seliones de terra clericorum de merton extendentes se in longitudine ultra le stoupendcrouchweye primi et occidentalissimi illius quarentene et iuxta viam sancti Neoti predictam et abuttant ad capud eorum boriale super foreram Johanis de Weston ultimo dictam *clar' hall*	 *iij. acr'* Egid.
Inter.ij. balcas	⎧ j.selio hospitalis predicti iuxta et abuttat super foreram predictam ⎨ iij.seliones cantarie beate marie ecclesie sepulcri iuxta abuttantes super ⎩ dictam foreram	*di acr'* Egid. *j acr'* Egid. 5a
♭ *Acr' et di*	[Loose insert: in late-sixteenth century hands:—] iiij curti Selliones quondam Thome Bole nuper Mr wood Iacentes in longitudine ad caput boriale quarti Selliones [sic] de Tumans Acre de terra Stephani Morris senioris qui est curtior aliis tribus Selionibus precedentes [sic] et iacent inter predictos curtiores Selliones hospitii predicti qui Iacet [sic] ex parte occidentale et ad caput borialle [sic] de prioris Dole qui vocatur Cuttinge ut supra in quarentena predicta ex una parte et predictos tres Selliones longiores de tumans Acre predicta ex parte altera et abbutant ad caput eorum boriale super foreram collegii predicti, quondam galfridi Seman, et balcantur ex utraque parte.	
	this note appeareth in all other terriers savinge in this of the Colledge *compared anno* 1594. *Eliz.* 36°	
[fol. 5v.] .*b*.	ij.seliones monialum de bech iuxta et abuttant super foreram predictam	*j acr'* Par.

GRYTHOWEFELD

stothydes acr'

++	mo	iiij.seliones Stephani morys senioris quondam patris sui qui vocantur tunmanis aker iuxta quorum.iij. primi seliones sunt longiores aliis selionibus antedictis magis quam in duplum et abuttant ad capud eorum boriale super foreram collegii predicti quondam Galfridi Seman / / et primus selio est in parte forera ad capud predictum et quartus selio est curtior aliis.iij.selionibus precedentibus et abuttat super terram Thome bolle ad capud predictum [—] Botulphi

cχc ♄   ij.seliones Collegii predicti quondam Galfridi Seman iuxta et abuttant
super terram dicti Thome Bolle ad capud eorum boriale   *iij rod b* Par.
j.selio hospitalis predicti iuxta et abuttat super terram dicti Thome
Bolle ad capud predictum   *di acr'* Egid.

cχc   j.selio collegii predicti quondam Galfridi Seman iuxta et est in parte
forera ad capud suum australe et abuttat ad capud suum boriale super
terram hospitalis predicti   *di [acr'] b* Par.
ij.seliones cum.j.curto selione qui vocatur a cuttyng iacentes in
longitudine ex parte occidentale ad capud boriale eorum de terra// hospitalis
predicti iuxta et sunt curtiores aliis selionibus antedictis ad capud
eorum australe et extendunt se in longitudine a forera Thome Morys
usque ad foreram Collegii predicti quondam Galfridi Seman ultimo
dictam et sunt primi et occidentalissimi dole dicti hospitalis que
vocatur prioresdole   *j acr' j rod.* Rotund'

nota   Et inter.vij^{um}. et.viij^{um}. seliones dictorum.xj.selionum iacet
Grythowweye
xj.seliones cum.j.gore iacentes ex parte orientale ad capud eorum
boriale eiusdem hospitalis iuxta quorum.iiij.seliones ultimi sunt
curtiores aliis selionibus antedictis ad capud predictum   *vj acr'* Egid.
v.seliones qui sunt longiores aliis selionibus antedictis ad capud eorum
boriale et.j.curtus selio qui vocatur le Cuttederode de terra Roberti
longe iuxta quorum primus selio est in parte forera ad capud predictum
   iij acr' Par.
j.selio continens.j.rode dicti Roberti Longe quondam Rogeri de
harleston et quondam Willelmi Ward et iacet in longitudine ad capud
boriale predicti cuttederod cuius decima   *di acr'* Egid.

	mo	⎧ iiij.seliones cum sua forera Stephani Morys senioris quondam Johanis berton iuxta	*iij acr'* Egid.
Inter.ij. balc'	mo	⎨ j.selio dicti Stephani quondam Johanis de Comberton iuxta	*di acr'* Rad.
cχ c		⎨ j.selio Collegii predicti quondam Thome de Cantebrigia iuxta	*j rod b* Rad.
		⎩ j.selio Roberti Tuliet iuxta nuper malbyrthorpe (deleted)	*di acr'* Rad.

[fol. 6r.]   6

		ij.seliones Prioris de Huntyngdon iuxta	*j acr'* Rad.
	mo	iiij.seliones cum sua forera Stephani Morys senioris quondam Johanis berton iuxta *di acr'* (deleted)	*iiij acr'* Egid.
	mo	j.selio dicti Stephani quondam Johanis de Redhod iuxta	*di acr'* Rad.
		j.selio prioris de huntyngdon et est longior aliis selionibus sequentibus in duplum iuxta	*di acr'* Rad.
		v.seliones hospitalis predicti iuxta	*j acr' di* Egid.
cχc		ij.seliones collegii predicti quondam Thome de Cantebrigia iuxta quorum ultimus selio est forera et sunt ultimi et orientalissimi illius quarentene *iij rod* (deleted) *b*	*j acr'* Rad.

95

THE WEST FIELDS OF CAMBRIDGE

| | *mo* | viij.seliones Thome Morys quorum orientalissimus selio est forera et iacent in longitudine ad capud boriale.ij.selionum Collegii predicti quondam Thome de Cantebrigia ultimo dictorum et sunt.j.quarentena per se | *ij acr' di* Egid. |

▯   7   Quarentena vocata brembilfurlang iacens transversa ad capud boriale.viij.selionum Thome Morys ultimo dictorum Seliones illius quarentene debent computari ad capud eorum orientale

cχ c   iij.seliones Collegii predicti quondam Galfridi Seman primi et australissimi illius quarentene quorum primus selio est forera
                                                                 ij acr' et di (deleted) 2 *acr'*ᵇ Par.

                ⌠j.selio hospitalis predicti iuxta   *j rod.* Egid.
Inter.ij. balcas ⟨ j.selio Thome marbilthorp quondam Willelmi lingwod iuxta   *clar' hall j rod* Par.
cχ c   ⌡j.selio Collegii predicti quondam Galfridi Seman iuxta   *di acr '*ᵇ Egid.
                 j.selio Capelle de Steresbreg iuxta   *di acr'* Par.
                 j.selio Rogeri de harleston quondam Ricardi Tableter iuxta   *di acr'* Par.
                 j.selio Ricardi Tuliet iuxta   *iij rod.* Rad.
                ⌠j.selio Thome bolle quondam Johanis de Toft iuxta   *di acr'* Rad
croft   ⌡j.selio Rogeri de Harleston quondam Ricardi Tableter iuxta   *di acr'* Par.
[fol. 6v.]   6a
Inter.ij. balcas ⟨ j.selio ocupatus per dictum Rogerum tanquam terre Roberti longe
                  iuxta   *di acr'* Par.
       o   ij.seliones dicti Rogeri quondam Ricardi Tableter iuxta   *j acr'.* Par.
clar' hall   ⌠j.selio Thome Marbilthorp quondam Willelmi Lingwod iuxta   *j rod di* Par.
Inter.ij. balcas ⌡j.selio hospitalis predicti iuxta   *j rod di* Egid.
                 j.selio Rogeri Harleston quondam Willelmi de lolleworth iuxta   *j rod* Par.
cχ c   iij.seliones Collegii predicti quondam Galfridi Seman iuxta et sunt
⊟   longiores aliis selionibus antedictis in duplum quorum primus selio est
                 in parte forera ad capud suum occidentale et tertius selio est in parte
                 forera ad capud suum orientale   *ij acre*   *ij acr'* Par.
cχ c   j.selio ocupatus per dictum Collegium iuxta et est curtior aliis
                 selionibus antedictis ad capud suum orientale et longior aliis
                 selionibus antedictis et in parte forera ad capud suum occidentale   ᵇ
                 di acra   *di acr'* Rad.
                 j.selio ocupatus per predictum hospitalem iuxta   *di acr'* Rad.
                 j.selio ocupatus per predictum hospitalem iuxta   *di acr'* Egid.
                 j.selio ocupatus per moniales sancti Radegunde iuxta   — Par.
cχ nota d   j.selio Collegii predicti vel hospitalis predicti quia in briga quondam
                 Galfridi Seman ut dicitur iuxta   *di acra* Par. ᵇ
                 j.selio ocupatus per capellam de Steresbreg' in cuius capite orientale
                 est wlwyesmere iuxta   *alias duckesmere*   *di acr'* Egid.
                 j.selio ocupatus per Rogerum de harleston tanquam terre Willelmi
                 Ward   *di acr'* Par.
                 j.selio ocupatus per hospitalem predictum iuxta   *di acr'* Egid.
                 j.selio ocupatus per capellam de Steresbreg' iuxta   *di acr'* Egid.
‡   iij.seliones Rogeri de Harleston quondam Willelmi de lolleworth
                 iuxta   *j acr' di* Par.
cχ c   j.selio Collegii predicti quondam Galfridi Seman iuxta et est longior
                 aliis selionibus antedictis et in parte forera ad capud suum orientale
                                                                                  iii rod. j acr' Par. ᵇ

GRYTHOWEFELD

   vij.seliones Stephani Morys senioris quondam patris sui iuxta et vocatur
   le nakededole/ quorum primus selio est curtior aliis selionibus sequentibus ad
   capud suum orientale et secundus selio est in parte forera ad capud
   predictum   *iiij acr'* Egid.
   ij.seliones Ricardi Tuliet iuxta   *j acr'* Rad.
   j.selio Rogeri de Harleston quondam Ricardi Tableter iuxta   *di acr'* Egid.

[fol. 7r.]   7
      iiij.seliones hospitalis predicti iuxta   *ij acr* Rad.
      ⎡j.selio Ricardi de londen quondam patris sui iuxta   *iij rod'* Egid.
clar' hall   ⎢j.selio dicti Ricardi quondam Johanis de Comberton iuxta   *j rod* Rad.
Intr.ij. balcas ⎨j.selio Collegii predicti quondam Thome de Cantebrigia iuxta   *j rod* Rad. *b*
cχ c   ⎣j.selio Rogeri de Harleston quondam Willelmi Warde iuxta   *j rod* Egid.
      ij.seliones Prioris de Huntyngdon iuxta   *j acr'* Rad.
   mo   viij.seliones Thome Morys iuxta   *iij acr'* Egid.
   mo   j.selio Stephani Morys senioris quondam patris sui iuxta   *j rod di* Egid.
      ⎡ij.seliones monialum sancte Radegunde iuxta   *j acr'* Par.
sine balca ⎨j.selio Roberti Longe iuxta et est ultimus et borialissimus illius
      ⎣quarentene et iuxta.ij.seliones collegii predicti quondam Galfridi
      Seman qui extendunt se in longitudine usque ad Huntyngdonweye
      prius computatur   *j acr'* Par.

     8   Quarentena parva iacens transversa ad capud orientale.iij.selionum
      Rogeri de Harleston ultimo dictorum in quarentena ultimo dicta
      ad tale signum   ‡ *incipe in parte australi*   *over thart the furlong of*
                                  duckmere
      ⎡j.selio cantarie beate marie ecclesie sancti sepulcri primus et
sine balca ⎨occidentalissimus illius quarentene et est forera   *di acr'* Egid.
      ⎢j.selio Roberti longe iuxta   *j rod* Par.
      ⎣ij.seliones Rogeri de harleston quondam Willelmi de lolleworth iuxta
                                j acr' Par.
      j.selio cantarie beate marie ecclesie Sancti Sepulcri iuxta et est longior
      aliis selionibus antedictis et in parte forera ad capud suum boriale   *j rod di* Egid.
      ij.seliones hospitalis predicti iuxta quorum ultimus selio est forera et sunt
      ultimi istius quarentene   *iij rod* Egid.

     9   Quarentena iacens transversa ad capud occidentale.iiij^{or}. selionum
      Collegii predicti quondam Galfridi Seman qui sunt primi   *ad talem*
      et australissimi quarentene de brembilfurlong antedicte *vel duckmere furlong*
      Seliones illius quarentene debent computari ad
      capud eorum australe.
                                       †

[fol. 7v.]
cχ c   ij.seliones Collegii predicti quondam Galfridi Seman qui vocantur le
      Gyldenaker et sunt primi et orientalissimi illius quarentene quorum
      primus selio est forera   *j acra* Par *b*.
   mo   ij.seliones Ricardi Morys iuxta   *j acr' di* Egid.
   mo   x.seliones eiusdem Ricardi iuxta quorum sextus selio est in parte
      forera ad capud suum australe et.iiij. ultimi seliones sunt curtiores
      aliis selionibus antedictis ad capud predictum   *vj acr'* Par.

97

THE WEST FIELDS OF CAMBRIDGE

      j.selio cantarie beate marie ecclesie sancti clementis quondam Ricardi
      Youn iuxta et est longior aliis selionibus antedictis et in parte forera
      ad capud boriale   *di acr'* Egid.
$c\chi$ c ⎧ j.selio Collegii predicti quondam Thome de Cantebrigia iuxta   *di acr'* Rad b
Inter.ij. balcas   ⎨ j.selio Ricardi Tuliet qui vocatur le Goredaker iuxta   ‡ *j acr'* Rad.
       j.selio Rogeri de harleston quondam Ricardi Tableter iuxta   *di acr'* Egid.
       j.selio hospitalis predicti iuxta   *di acr'* Egid.
mo   ij.seliones Thome Morys iuxta quorum ultimus selio est forera et sunt
       ultimi et occidentalissimi illius quarentene   *j acr'* Egid.

♄   10   Quarentena curta continens.viij.buttes iacens transversa ad capud
       australe selionis Collegii predicti quondam Thome de Cantebrigia
       ultimo dicti et abuttat ad capud suum occidentale super Grythowweye
       Incipe in parte occidentali   ‡
       ⎧ ij.buttes Rogeri de Harleston quondam Ricardi Tableter prime et
sine balca   ⎨ borialissime illius quarentene quarum prima butta est forera   *di acr'* Egid.
       ⎩ ij.buttes Stephani Morys quondam patris sui iuxta   *di acr'* Egid.
       iiij.buttes hospitalis predicti iuxta quarum ultima butta est forera
       et sunt ultime et australissime illius quarentene   *di acr'* Egid.

♄   11   Quarentena curta continens.vij.buttes abuttans directe ad capud suum

[fol. 8r.]   8

       orientale contra capud occidentale quarentene curte ultime dicte
       mediante Grythowweye predicta
$c\chi$ c ⎧ j.butta Collegii predicti quondam Galfridi Seman prima et australissima
sine balca   ⎨ illius quarentene et est forera   *di acr'* Par. b
       ⎩ ij.buttes hospitalis predicti iuxta   *di acr'* Egid.
mo   ⎧ ij.buttes Stephani Morys senioris quondam patris sui iuxta   *di acr'* Egid.
       ⎩ ij.buttes Rogeri de Harleston quondam Ricardi Tableter iuxta
       et sunt ultime et borialissime illius quarentene et iuxta.ij.seliones
       Stephani Morys junioris quondam Galfridi Seman qui extendunt se
       in longitudine a Grythowweye predicta usque ad viam sancti Neoti
                           di acr' Egid.
       iiij.curti seliones Thome bolle iacentes in longitudine ad capud boriale
       quarti selionis de Tunmannisaker de terra Stephani morys senioris
       qui est curtior aliis.iij.selionibus precedentibus ut supradictum   ++ *stacked acr'*
       est in quarentena de Stoupendecrouch predicta et iacent
       inter predictum curtum selionem dicti hospitalis qui iacet ex parte
       occidentali ad capud boriale de priourisdole et vocatur a cuttyng ut
       supra in quarentena predicta ex una parte et predictos.iij.seliones
       longos de Tunmannisaker predicta ex altera parte et abuttant ad
       capud eorum boriale super foreram Collegii predicti quondam   ++
       Galfridi Seman ultimo dictam et balcati ex utraque parte   *sunt quarentena*
                           per se ij acr' Rad.
mer'   v.buttes de terra clericorum de Merton in quibus sunt le gravelpittis
       iacentes in longitudine iuxta Grythowepath predictam et abuttant ad
       capud eorum australe super viam sancti neoti predictam et sunt.j.
       quarentena per se *hii reliqui seliones non possunt distingui quia non apparet*
                        quod solvuntur. Egid.

GRYTHOWEFELD

Grythowe
⦀⦀ 🕯 *12*   Quarentena in qua est Grythowehill iacens transversa ad capud
boriale.v.buttes de terra clericorum de merton ultimo dictarum et
abuttes super Grythowepath / / Seliones illius quarentene debent
computari ad capud eorum orientale
*sed reliqui seliones incipere per le condyte incipiendo in fine posteriore
quarentene predicte* [- - - -] *ad tale signum* △ *ex altera pagina*

[fol. 8v.]

	⎧ j.selio Thome atte cherch de howes quondam Willelmi de lavenham	
	⎪ primus et australissimus illius quarentene et est forera	*di acr'* Rad.
sine balca	⎨ ij.seliones Prioris de huntyngdon iuxta	*j acr'* Rad.
mer'	⎩ ij.seliones de terra clericorum de merton iuxta	*j acr'* Egid.
	ij.seliones cantarie beate marie ecclesie sancti sepulcri iuxta	*j acr'* Egid.
.b.	j.selio monialum de bech iuxta	*iij rod* Par.
mo	j.selio Thome morys iuxta	*iij rod* Egid.
mer	v.seliones clericorum de merton iuxta et in capite orientali ultimi	
	selionis est Grythowehill	*ij acr' di* Egid.
cχ c	ij.seliones Collegii predicti quondam Galfridi Seman iuxta	*j acr' et di* Par. *b*
	ij.seliones cantarie beate marie ecclesie sancti sepulcri iuxta	*j acr'* Egid.
cχ c	j.selio Collegii predicti quondam Thome de Cantebrigia iuxta	*di acr'* Rad. *b*
hospitalis	j.selio Thome atte cherch de howes quondam Willelmi lavenham iuxta	
		di acr' Rad.
mo	ij.seliones Thome morys iuxta	*iij rod* Egid.
	ij.seliones qui solebant esse.iv.cum.j.forera Prioris de Bernewell	
△	iuxta et sunt curtiores aliis selionibus antedictis ad capud eorum	
	orientale et sunt ultimi et borialissimi illius quarentene et iuxta le	
Cundytehed	cundut	*j acr'* Egid.
	iiij.seliones et.j.gora cum suis foreris Rogeri de harleston quondam	
	Galfridi Seman iacentes transversi ad capud occidentale quarentene	
	ultimo dicte et iacent in longitudine iuxta foveam vocatam Semanis	
	dich quorum.ij.orientales seliones sunt longiores aliis.ij.selionibus ad	
peperdole	capud eorum australe et orientalis selio est forera et sunt.j.quarentena	
	per se	*iij acr'* Par.

🕯 *13*   Quarentena iacens in longitudine inter melneweye et kynchmade
Seliones illius quarentene debent computari ad capud eorum orientale
mer'   iiij.buttes de terra clericorum de merton prime et australissime illius
quarentene   *di acr'* Egid.
iij.buttes Rogeri de harleston quondam Ricardi Tableter iuxta   *di acr'* Egid.

9

[fol. 9r.]

	ix.buttes Prioris de Bernwell iuxta	*ij acr' j rod* Egid.
	v.seliones hospitalis predicti iuxta	*j acr' j rod* Rad.
§ cχ c	iij.seliones Collegii predicti quondam Galfridi Seman iuxta	*ij acr'* Par. *b*
	j.selio cantarie beate marie ecclesie sancti sepulcri iuxta	*j acr'* Egid.
.b.	j.selio monialum de bech iuxta	*j acr'* Par.
	ij.seliones Roberti Longe iuxta *bokenham rawlyn*	*j acr' j rod* Par.
§ cχ c	⎧ j.selio Collegii predicti quondam Thome de Cantebrigia iuxta et est	
	⎨ curtior aliis selionibus antedictis ad capud suum occidentale	*di acra*
Inter.ij. balcas	⎨	*di acr'* Rad' *b*
mo	⎩ iiij.seliones Stephani Morys senioris quondam patris sui iuxta	*j acr' di* Egid.

99

THF WEST FIELDS OF CAMBRIDGE

Inter.ij. balcas mo	⎧ iiij.seliones hospitalis predicti iuxta ⎨ iiij.seliones Rogeri de harleston quondam Ricardi Tableter iuxta ⎩ j.selio Stephani Morys senioris quondam Johanis de Comberton iuxta	*ij acr'* Par. *ij acr'* Par. *di acr'* Rad.
§ cχ c Inter.ij. balcas mo mer'	⎧ j.selio Collegii predicti quondam Thome de Cantebrigia iuxta ⎨ ⎩ j.selio Thome Morys iuxta v.seliones de terra clericorum de merton iuxta j.selio Rogeri de harleston quondam Willelmi Ward iuxta ij.seliones Prioris de Bernewell iuxta	*di acr'* Rad. *j acr'* Egid. *ij acr' di* Rad. *j acr'* Egid. *di acr'* Egid.
mo	j.selio Stephani morys senioris quondam Johanis Redhod iuxta	*j acr'* Par.
§ cχc Inter.ij. balcas mo mo mo sine balca	⎧ j.selio Thome Jekke quondam Johanis blancpayn iuxta ⎨ j.selio Collegii predicti quondam Thome de Cantebrigia iuxta ⎨ ⎨ j.selio Thome Morys iuxta ⎨ j.selio Stephani Morys senioris quondam Rogeri Houdlo iuxta ⎨ ij.seliones dicti Stephani quondam patris sui ⎨ ij.seliones Prioris de bernewell iuxta quorum ultimus selio est forera ⎩ et sunt ultimi et borialissimi illius quarentene	*di acr'* Rad. *j rod* Rad. *j acr'* Egid. *j acr'* Par. *iij rod* Egid. *iij rod* Egid.

☞ 14 Quarentena abuttans ad capud suum australe super foreram Prioris de bernewell ultimo dictam et ad capud suum boriale super Gryttonfeld.

[fol. 9v.]

ecclesie marie in foro	⎧ iij.seliones qui solebant esse.vj.Thome atte cherch de Howes quondam ⎨ Willelmi de Lavenham et quondam Ricardi youn primi et orientalissimi ⎨ et primus selio est forera .ij.acr. ⎨ j.selio eiusdem Thome quondam Simonis houdlo de howes iuxta ⎨ j.selio dicti Thome quondam Willelmi de Lavenham et quondam ⎨ Ricardi youn iuxta iij. Rod. ⎩ j.selio Thome Jekke quondam Johanis blancpayn iuxta. iij. Rod xij.seliones quondam Roberti Dunnyng iuxta et ultimi et iuxta kynchmad iij. acr' *iste vocatur spaldynges closs nunc gravelpyttes.*	Egid. iij Rod Rad. Rad. Rad. Egid.

☞ 15 Quarentena iacens in longitudine inter kynchmad et braderussh et abuttat ad capud suum occidentale super braderussh / / Seliones illius quarentene debent computari ad capud eorum occidentale

	ix.seliones hospitalis predicti qui vocantur le mordole iacentes in longitudine iuxta campum de girton quadam lata balca mediante et sunt primi et borialissimi illius quarentene	*vj acr' di* Egid.
mo	vj.seliones Thome Morys iuxta	*iij acr'* Egid.
	iiij.seliones Rogeri de Harleston quondam Roberti goldsmyth iuxta	*ij acr* Par.
	xj.seliones de terra mortimer iuxta	*vj acr'.* Rad.
	viij.seliones et.j.gora Prioris de bernewell iuxta	*iiij acr'* Egid.
cχ ca v rods	ij.seliones Collegii predicti quondam Thome de Cantebrigia iuxta	*j acra j rod* Rad.
Inter.ij. balcas	⎧ ix.seliones Prioris de huntyngdon iuxta ⎩ ij.seliones Ricardi Niket quondam Johanis de Refham iuxta	*iij acr* Rad. *j acr'* Rad.

ecclesie marie in foro

GRYTHOWEFELD

mo	ij.seliones Thome Morys iuxta	*ij acr'* Egid.
	iij.seliones Rogeri de Harleston quondam Ricardi Tableter iuxta	*iij rod* Rad.
	viij.seliones cum sua forera Prioris de bernewell iuxta	*iiij acr'* Rad.
mo	xxiiij.seliones cum sua forera Stephani Morys senioris quondam Johanis de Berton iuxta et sunt ultimi et australissimi illius quarentene et prope le cundut	*ix acr'* Egid.

10

[fol. 10r.]

☞ *16* Quarentena ex altera parte de braderussh predicto abuttans ad capud suum orientale super bradrussh predictum ex oposito quarentene ultimo dicte // Seliones illius quarentene debent computari ad capud eorum orientale

mo sine balca	⎧ vj.buttes Stephani morys junioris quondam Johanis Pittok primi et australissimi illius quarentene et abuttant ad capud eorum orientale super bradrussh predictum et ad capud eorum occidentale super viam sancti neoti ex oposito Semanis dich predicti	*j acr'* Egid.
	⎩ vj.seliones Prioris de bernwell iuxta et abuttant ut supra	*j acr' di* Egid.
o	⎧ ij.seliones Roberti longe vel Galfridi Seman quia in briga et placito iuxta et abuttant ut supra *jankin rawlyns bokenham*	*j acr'* Par.
Inter.ij. balcas	⎨ iij.seliones Prioris de Bernewell iuxta et abuttant ut supra	*j acr'* Egid.
	⎩ iij.seliones monialum Sancte Radegunde iuxta et abuttant ut supra	*j acr' di* Par.
mo	ij.seliones Stephani Morys junioris quondam Johanis Pittok iuxta et abuttant ut supra	*j acr'* Par.
mo	j.selio dicti Stephani quondam Johanis de Comberton iuxta et abuttat ut supra	*di acr'* Rad.
cχc	iiij.seliones Collegii predicti quondam Thome de Cantebrigia iuxta quorum.iij.primi seliones abuttant ut supra et quartus selio est curtior aliis selionibus antedictis ad capud suum orientale et abuttat super foreram Prioris de Bernewell quondam Roberti longe ad capud predictum	*ij acra et di b* Rad.
	iiij.seliones Prioris de Bernewell iuxta et abuttant super foreram predictam	*ij acr'* Egid.
mo Inter.ij. balcas	⎧ j.selio Stephani morys senioris quondam Willelmi de Bekeswell iuxta et abuttat super foreram predictam	*di acr'* Par.
	⎩ j.selio Rogeri de Harleston quondam Willelmi de lolleworth iuxta et abuttat super foreram predictam	*di acr'* Par.
.b.	⎧ j.selio monialum de bech iuxta et abuttat super foreram predictam	*j rod di* Par.
Inter.ij. balcas	⎩ j.selio Prioris de Bernewell iuxta et abuttat super foreram predictam	*j rod* Egid.
mo	j.selio Stephani Morys senioris quondam patris sui iuxta et abuttat super foreram predictam	*j acr'* Egid.
.b.	ij.seliones monialum de bech iuxta et abuttat super foreram predictam	*j acr' di* Par.
	ij.seliones monialum Sancte Radegunde iuxta et abuttant super foreram predictam	*ij acr'* Par.

[fol. 10v.]

cχ ca	ij.seliones Collegii predicti quondam Thome de Cantebrigia iuxta et abuttant super foreram predictam *j acra et di* (deleted)	2 *acr'* Rad.

Memorandum est quod omnes seliones predicti abuttant super viam sancti neoti antedictam ad capud eorum occidentale

THE WEST FIELDS OF CAMBRIDGE

Inter.ij. balcas ✳	⎧ iiij.seliones cantarie beate marie ecclesie sancti sepulcri iuxta et sunt curtiores aliis selionibus antedictis ad capud eorum occidentale et abuttant super foreram predictam	*j acr' di* Egid.
	⎩ ij.seliones Cantarie ecclesie clementis iuxta et abuttant super foreram predictam	*j acr' di* Par.
f cχ	j.selio Collegii predicti quondam Galfridi Seman iuxta abuttans super foreram predictam	*di acr'* Egid.
	iij.seliones Cantarie ecclesie sepulcri iuxta et sunt longiores aliis selionibus antedictis ad capud eorum orientale et abuttant super braderussh predictum quorum primus selio est in parte forera ad capud predictum	*j acr' di* Egid.
k.h.	⎧ ij.seliones cum.j.gora ex parte boriali ad capud eorum occidentale Roberti longe iuxta et vocatur le Goredaker	Par.
	⎩ vj.seliones Prioris de Bernewell iuxta	*j acr' iij rod* Egid.
Inter.ij. balcas	⎧ j.selio hospitalis predicti iuxta	*j rod* Egid.
	⎨ j.selio monalium Sancte Radegunde iuxta	*j rod* Par.
	⎩ j.selio Hospitalis predicti iuxta	*j rod* Egid.
	v.seliones Prioris de Bernewell iuxta	*j acr' di* Egid.
	j.selio hospitalis predicti iuxta	*iij rod* Egid.
Inter.ij. balcas ✳	⎧ j.selio Rogeri harleston quondam Roberti Goldsmyth iuxta	*j rod* Rad.
	⎨ ij.seliones Prioris de huntyngdon iuxta	*iij rod* Rad.
	⎩ iiij.seliones Johanis Cotten quondam Thome de Comberton iuxta	*j acr'* Rad.
Inter.ij. balcas	⎧ iij.seliones hospitalis predicti iuxta et sunt longiores aliis selionibus antedictis ad capud eorum occidentale et primus selio est in parte forera ad capud predictum	*j acr' di* Egid.
	⎩ j.selio cantarie ecclesie sepulcri iuxta	*di acr'* Egid.
mer'	ij.seliones de terra clericorum de merton iuxta	*iij rod* Egid.
	iij.seliones Rogeri de Harleston quondam Willelmi Ward iuxta quorum ultimus selio est forera et sunt ultimi et borialissimi illius quarentene	*j acr'* Egid.
.b.	ij.seliones cantarie ecclesie clementis ultimo dictorum	
†	ad tale signum (whole entry deleted) ✳	

11

[fol. 11r.]

♭	17	Quarentena continens.iij.seliones et iacet transversa ad capud orientale	
†		*ij selionum cantarie clementis predictorum ad talem* [sic] *signum* ✳	
		ij.seliones Prioris de bernewell quondam Roberti longe quorum occidentalis selio est forera	*j acr'* Par.
bokenham		j.selio dicti Roberti longe iuxta et iacet in longitudine iuxta braderussh predictum	*di acr'* Par.
♭	18	Quarentena iacens in longitudine iuxta bradrussh predictum et abuttat ad capud suum boriale super moram de Maddingle / / Seliones illius quarentene debent computari ad capud eorum boriale	
		xiiij.seliones cum sua forera Prioris de Bernewell primi et orientalissimi illius quarentene iuxta braderussh predictum et abuttant ad capud eorum australe super foreram Rogeri de harleston ultimo dictam et ad capud eorum boriale super moram de Maddingle predictam	*iij acr' di* Egid.

GRYTHOWEFELD

Inter.ij. balcas	ix.seliones cum sua forera Rogeri de harleston quondam Willelmi Ward iuxta et abuttant ad capud eorum boriale super moram predictam quorum quintus selio est in parte forera ad capud suum australe et.iiij.ultimi seliones sunt curtiores aliis selionibus antedictis ad capud predictum	*ij acr'* Par.
	vij.seliones Prioris de Bernewell iuxta et abuttant super moram predictam	*j acr' iij rod* Egid.
*ca c*χ	vj.seliones Collegii predicti quondam Thome de Cantebrigia iuxta	
ij acr j rod	quorum.iiij.primi seliones abuttant super moram predictam et. iij.ultimi seliones abuttant super foreram Prioris de bernewell et ultimus selio est forera et sunt ultimi et occident[al]issimi illius quarentene	*ij acr'* Rad.

19 Quarentena iacens transversa ad capud occidentale.iiij.ᵒʳ selionum Johanis de Cotten ultimo dictorum ad tale signum ✷ Seliones illius quarentene debent computari ad capud eorum boriale

[fol. 11v.]

	iiij.seliones Johanis de Cotton quondam Thome de Comberton primi et orientalissimi illius quarentene quorum.j.selio est forera	*j acr'* Rad.
*ca c*χ	iij.seliones Collegii predicti quondam Thome de Cantebrigia iuxta quorum primus selio est curtior aliis selionibus sequentibus ad capud suum	
Inter.ij. balcas	boriale et medius selio est in parte forera ad capud predictum	*iij rod* Rad.
	ij.seliones Prioris de Bernewell iuxta	*iij rod* Egid.
mo	vj.seliones Thome Morys iuxta	*j acr' di* Egid.
*ca c*χ	j.selio Collegii predicti quondam Thome de Cantebrigia iuxta	*j rod* Rad.
Inter.ij.balcas	j.selio Johanis Norton quondam Johanis de Comberton iuxta	*j rod* Rad.
	iiij.seliones Prioris de Bernewell iuxta	*j acr'* Egid.
mer'	v.gores et.j.selio cum una butta iacentes ex parte orientali ad capud eorum boriale de terra clericorum de merton iuxta et sunt longiores	
sine balca	aliis selionibus antedictis ad capud predictum et predicta butta est forera et iacet in longitudine ad capud boriale quarti selionis.iiijᵒʳ. selionum Prioris de bernewell ultimo dictorum	*j acr'* Egid.
*f c*χ	j.selio Collegii predicti quondam Thome de Cantebrigia iuxta	*j rod* Rad.
	j.selio Johanis Norton quondam Johanis de Comberton iuxta	*j rod* Rad.
mer'	v.seliones de terra clericorum de merton iuxta et sunt ultimi et occidentalissimi illius quarentene et iuxta viam Sancti Neoti	*ij acr'* Egid.

20 Quarentena iacens transversa ad capud australe.v.selionum clericorum de merton ultimo dictorum et abuttat ad capud suum occidentale super viam sancti Neoti / / Seliones illius quarentene debent computari ad capud eorum occidentale

	iiij.seliones Prioris de Bernewell **primi** et borialissimi illius quarentene et primus selio est forera	*ij acr'* Egid.
	j.selio Johanis de Cotton quondam Thome de Comberton iuxta	*di acr'* Rad.
*ca c*χ	j.selio Collegii predicti quondam Galfridi Seman iuxta	*di acr'* Egid.
Inter.ij. balcas	ij.gores et.iij.seliones qui solebant esse.viij. Johanis de Cotten quondam Thome de Comberton iuxta	*ij acr'* Rad.
	iiij.seliones Prioris de Bernewell iuxta	*j acr'* Egid.
.f		

THE WEST FIELDS OF CAMBRIDGE

[fol. 12r.]

Inter.ij. balcas	ij.seliones Cantarie beate marie ecclesie sancti Clementis iuxta	j acr' Par.
	iiij.seliones Cantarie beate marie ecclesie sancti sepulcri iuxta	j acr' Egid.
	ij.seliones ocupati per Priorem de Bernewell iuxta	di acr' Rad.
mo	ij.seliones Stephani Morys senioris quondam patris sui iuxta	ij rod Egid.
b ca cχ	iij.seliones Collegii predicti quondam Thome de Cantebrigia iuxta et sunt ultimi et australissimi illius quarentene et iuxta.ij.longos seliones dicti Collegii quondam dicti Thome extendentes se in longitudine a via sancti Neoti predicta usque ad foreram Prioris de Bernewell quondam Roberti longe / /	iij Rod' Rad.

♗ 21   Quarentena curta continens.xx.seliones iacens transversa ad capud boriale.v.selionum clericorum de merton ultimo dictorum in quarentena .ij.precedente et abuttat ad capud suum occidentale super viam sancti
✠   neoti antedictam ex oposito alte cruci [sic]

sine balca	xiij.seliones Prioris de Bernewell primi et australissimi illius quarentene quorum primus selio est forera	iij acr' j rod Egid.
	vij.seliones eiusdem Prioris iuxta quorum ultimus selio est forera et sunt ultimi et borialissimi illius quarentene iij (deleted) j acr' iij rod Botul.	

nota

♗ 22   Quarentena vocata le chekker iacens in longitudine ad capud orientale quarentene curte ultimo dicte / / Seliones illius quarentene Debent computari ad capud eorum occidentale

sine balca	v.seliones Prioris de Bernewell primi et australissimi illius quarentene et abuttant ad capud eorum occidentale super buttam clericorum de merton ultimo dictam que est forera ut supradictum est in quarentena .iiij.precedente quorum primus selio est forera	j acr' j rod Egid.
	ij.seliones Johanis Norton quondam Johanis blancpayn iuxta et abuttant ut supra	di acr' Par.
cχ ca	j.selio Collegii predicti quondam Galfridi Seman iuxta et est longior aliis selionibus ad capud orientale	iij rod j acr' Egid.
Inter.ij. balc'	j.selio hospitalis predicti iuxta et est in parte forera ad capud suum orientale	di acr' Egid.

[fol. 12v.]

	xiiij.seliones cum una gora iacentes ex parte australi ad capud eorum orientale Thome de Audele iuxta et sunt curtiores aliis selionibus antedictis ad capud predictum	v acr' j rod Par.
	xij.seliones Prioris de Bernwell iuxta quorum.iij.ultimi seliones sunt longiores aliis selionibus antedictis ad capud eorum orientale et decimus selio est in parte forera ad capud predictum	vj acr' Egid.
mo	iij.seliones Stephani morys senioris quondam Rogeri houdlo iuxta	j acr' Rad.
sine balca	ij.seliones Collegii predicti quondam Thome de Cantebrigia iuxta	iij rod Rad.
ca cχ	xiiij.seliones Prioris de Bernewell iuxta quorum.xj.seliones ultimi habent suam foreram propriam ad capud eorum orientale et sunt ultimi et borialissimi illius quarentene	vij acr' Rad.

MIDDELFELD

☝ 23 Quarentena iacens transversa ad capud occidentale.xiiij.selionum
Prioris de Bernewell ultimo dictorum / / Seliones illius quarentene
debent computari ad capud eorum boriale

ca cχ ⎧ viij.seliones Prioris de Bernewell primi et orientalissimi illius quarentene
sine balca ⎨ quorum primus selio est forera   *iij acr'* Egid.
         ⎩ ij.seliones Collegii predicti quondam Thome de Cantebrigia iuxta *iij rod* Rad.
ij.seliones Henrici blancpayn de Gyrton quondam patris sui iuxta *j acr'* Rad.
vj.seliones Prioris de Bernewell iuxta et sunt ultimi et occidentalissimi
illius quarentene et iuxta viam Sancti Neoti   *iij acr'* Egid.

☝ 24 Quarentena iacens transversa ad capud boriale quarentene ultimo
dicte et abuttat ad unum capud super viam sancti Neoti et ad aliud
capud super moram de maddingle

          ⎧ xvj.seliones et.j.gora cum sua forera Prioris de bernewell primi
sine balca ⎨ et australissimi illius quarentene et primus selio est forera *ix acr* Egid.
          ⎩ iiij.seliones Thome de Audle iuxta   *j acr di* Par.
ij.seliones Rogeri Baldiston quondam Johanis de Comberton
iuxta   *j acr'* Rad.
ca cχ j acr ⎧ ij.seliones Collegii predicti quondam Thome de Cantebrigia iuxta *j acr'* Rad.
Inter.ij. balc' ⎩ iiij.seliones Prioris de Bernewell iuxta   *j acr'* Rad.
[here a footnote intended for the end of Grithowefeld. see fol. 13r.]

[fol. 13r.]   13

ix.seliones Prioris de huntyngdon iuxta   *iij acr'* Rad.
viij.seliones tum aliqui eorum sunt gores Prioris de Bernewell iuxta
quadam lata balca mediante et sunt ultimi et borialissimi illius
quarentene et prope le morbernes   *j acr' di* Egid.
Et sic finitur Grythowefeld
                          | .ix. |

[footnote] *Summa selionum istius campi decimalium ecclesie botulphi xv b xj*

Middelfeld

Memorandum quod garbe decimalium de omnibus selionibus
iacentibus in middelfeld qui dicuntur.par. ut infrascriptum est debent
eque partiri inter Radegundam et Egidium videlicet una garba
Radegunde et.j.garba Egidio
                                                  saynt Johns closse
☝ *I.23* Quarentena vocata muscroftfurlong abuttans super capud suum
occidentale super viam sancti Neoti *Muscrofte ys the closse on the north
parte of seynt Johns bernys the lane betwyx.*

[fol. 13v.]   1b

hospitalis xx.seliones cum suis foreris quondam Roberti Dunnyng qui vocantur
muscroft primi et australissimi illius quarentene et prope seint iones
bernes   *xj acr* Egid.
iiij.seliones Ricardi Tuliet iuxta mediante bertonweye   *ij acr* Rad.
iiij.seliones Thome Boll quondam Johanis de Toft iuxta   *ij acr* Rad.
:+:   ij.seliones Stephani Morys senioris quondam Johanis blangron
mo   pelliparii iuxta   *j acr* Egid.

THE WEST FIELDS OF CAMBRIDGE

gun(vile)	iiij.seliones de terra Mortimer iuxta	*ij acr* Rad.
	v.seliones hospitalis sancti Johanis Evangeliste Cantebrigiensis iuxta	*iij acr* Egid.
gun(vile)	iij.seliones de terra Mortimer iuxta	*j acr* Rad.
	iiij.seliones hospitalis predicti iuxta	*ij acr* Egid.
gun(vile)	iij.seliones de terra Mortimer iuxta	*j acr* Rad.
	v.seliones hospitalis predicti iuxta	*ij acr di* Rad.
	iij.seliones Rogeri de Harleston quondam Ricardi Tableter iuxta	*j acr di* Rad.
mo	j.selio Stephani Moriz senioris quondam Johanis Redhod iuxta	*di acr* Rad.
	j.selio Prioris de Huntyngdon iuxta	*ij rod* Rad.
	j.selio Roberti longe iuxta	*iij rod* Par.
cχ	⎧ j.selio hospitalis predicti iuxta	*di acr* Egid.
Inter ij. balcas	⎨ ij.seliones corporis christi quondam Galfridi Seman iuxta *j acra*	*j rod b* Par.
b	⎩ ij.seliones monialum de bech iuxta	*di acr* Par.
	j.selio Prioris de huntyngdon iuxta	*di acr* Rad.
	vj.seliones hospitalis predicti iuxta	*j acr di* Egid.
mo	vj.seliones Stephani Moriz senioris quondam Johanis de berton iuxta quorum ultimus selio est forera et sunt ultimi et boriales illius quarentene	*j acr di* Egid.
gun(vile) et q.	viij.seliones cum sua forera de terra Mortimer qui vocantur chalkwell dole iacentes transversi ad capud orientale.iij.selionum Thome bolle ultimo dictorum in quarentene ultimo dicte et sunt una quarentena per se	*iij acr* Rad. :+:
hospitalis et q.	iiij.seliones cum suis foreris Thome bolle quondam Johanis de Toft qui vocantur frosshiscroft iacentes ex altera parte vie sancti	*j acr di* Rad.
Froshyscrofte nunc vocatur hunellys close	Neoti predicte inter dictam viam ex una parte et Wlwyes dich ex altera parte et abuttant ad capud eorum boriale super le brook ad le gravelpit et sunt una quarentena per se	

Chalkewell the sprynge att the north weste corner of clare hall ferme

[fol. 14r.]   14

2 ♭ 24		Quarentena iacens transversa ad capud australe iiij^{or}.selionum Thome Bolle qui vocantur Froshis croft ultimo dictorum abuttans super dictam viam Neoti *versus orientem.* *Incipe in parte boriale*	
	mo	ix.seliones qui solebant esse.x.Stephani Moriz senioris quondam patris sui primi et borialissimi illius quarentene et abuttant ad capud eorum orientale super viam sancti Neoti antedictam et ad capud eorum occidentale super Wlwyes dich predictum	*sine balca* *ij acr* Egid.
clar' hall		ij.seliones Johanis de Weston quondam hugonis Pittok iuxta	*di acr* Egid.
	mo	ij.seliones Stephani Moriz senioris quondam patris sui iuxta	*di acr* Egid.
	mo	j.selio Stephani Moriz junioris quondam Johanis Pittok iuxta	*j rod* Par.
	mo	ij.seliones Stephani Moriz senioris quondam patris sui iuxta	*di acr* Par.
	b	ij.seliones monialum de bech iuxta	*iij rod* Par.
		⎧ iij.seliones hospitalis predicti iuxta	*iij rod* Egid.
Inter.ij. balcas		⎨ j.selio Rogeri de harleston quondam Willelmi de lolleworth iuxta	*j rod* Par.
b Inter.ij. balcas		⎩ iiij.seliones sancti Egidii qui vocantur seyntgilis aker iuxta	*j acr* Par.
cχ		⎣ j.selio Collegii predicti quondam Galfridi Seman iuxta	*j rod b* Par.
.§.*c*		ij.seliones hospitalis predicti iuxta	*di acr* Rad.

MIDDELFELD

hospitalis Inter.ij. balcas	⎧ j.selio Stephani Moriz senioris quondam Willelmi de bekeswell iuxta		*j rod* Par.
	⎩ v.seliones hospitalis predicti iuxta		*v rod* Egid.
clar' hall Inter.ij. balcas *cχ c*	⎧ j.selio Thome marbilthorp iuxta		*j rod* Par.
	⎩ j.selio Collegii predicti quondam Galfridi Seman iuxta		*j rod b* Par.
b	iij.seliones monialum de bech iuxta		*j acr* Par.
mo	iij.seliones Stephani Moriz senioris quondam patris sui iuxta		*iij rod* Egid.
mo	iij.seliones dicti Stephani quondam domini Nicholai vicarii ecclesie clementis iuxta		*j acr* Par.
mo	ij.seliones dicti Stephani quondam Johanis Redhod iuxta		*j acr* Par.
	iij.seliones hospitalis predicti iuxta		*j acr* Egid.
b	⎧ iij.seliones monialium de bech iuxta		*j acr* Par.
Inter.ij. balcas *cχ c*	⎨ ij.seliones Collegii predicti quondam Thome de Cantebrigia iuxta		
	⎩		*iij rod b* Rad.
	ij.seliones hospitalis predicti iuxta		*di acr* Egid.

2b

[fol. 14v.]

mo	x.seliones Stephani Moriz senioris quondam Johanis de berton iuxta cum sua forera		*v acr* Egid.
	iiij.seliones Prioris de huntyngdon iuxta cum sua forera		*ij acr* Rad.
b	j.selio monialum de bech iuxta		*di acr* Par.
mo	⎧ j.selio Thome Morys iuxta et est curtior aliis selionibus antedictis		
Inter.ij. balcas	⎨ ad capud occidentale iuxta		*di acr* Egid.
mo	⎩ j.selio Stephani Moriz senioris quondam Johanis de Comberton iuxta		*di acr* Rad.
	vij.seliones Roberti de Brigham iuxta		*ij acr* Egid.
	⎧ vj.seliones Prioris de huntyngdon iuxta		*iij acr* Rad.
mo	⎨ v.seliones Thome Moriz iuxta		*ij acr* Egid.
mo sine balca	⎩ j.selio Stephani Moriz senioris quondam Johanis Redhod et est ultimus et australissimus illius quarentene et est forera		*di acr* Rad.

shepcote

3 24	Quarentena in qua situata est bercaria hospitalis predicti et abuttat ad capud suum australe super bertonweye / / Seliones illius quarentene debent computari ad capud eorum australe		
a	vj.seliones Prioris de Bernewell primi et orientalissimi illius quarentene et iuxta viam sancti Neoti antedictam et abuttant ad capud eorum australe super berton Weye predictam et ad capud eorum boriale super foreram Stephani moris senioris ultimo dictam		*iij acr'* Egid.
Inter.ij. balcas	⎧ j.selio Rogeri de harleston quondam Willelmi Ward iuxta		*di acr'* Egid.
lovel mo	⎪ j.selio hospitalis predicti iuxta		*j rod* Egid.
Inter.ij. balcas	⎨ ij.seliones Stephani Moriz (sen *deleted*) junioris quondam Willelmi Tuliet et postea quondam Pittok iuxta		*iij rod* Rad.
	⎪ j.selio Rogeri de harleston quondam Ricardi Tableter iuxta		*j rod* Egid.
gylys rode	⎩ j.selio Willelmi Burton quondam Willelmi Dene Skynnere iuxta		*j rod* Egid.
	v.seliones hospitalis predicti iuxta		*iij acr* Egid.
Hunnell crosse ✠	vj.seliones Rogeri de harleston quondam Ricardi Tableter iuxta.		*iij acr* Rad.
in berton weye	x.seliones et.j.gora prioris de huntyngdon iuxta		*iij acr* Rad.

THE WEST FIELDS OF CAMBRIDGE

ex opposito   Seynt Johnis berkarye or shepecotte ys bye the ix or x welowes on the
collegio Johanis   north syde of berton weye by small mede weste from Hunnell crosse
in berton weye forseyd.

15

[fol. 15r.]

$c"$ $c\chi$		v.seliones Collegii predicti quondam Thome de Cantebrigia		*b*
Inter.ij. balcas		iuxta *j acra*		*iij rod* Rad.
		j.selio hospitalis predicti iuxta		*j rod di* Egid.
clar' hall		iiij.seliones Johanis de Weston quondam Johanis de Comberton		
		iuxta		*iij rod* Rad.
mo		ij.seliones Stephani Moriz senioris quondam Johanis Redhod		
		iuxta		*j acr'* Par.
sine balca		iiij.seliones hospitalis predicti iuxta et in capite australi ultimi		
		selionis situatur bercaria hospitalis predicti et sunt ultimi et		
		occidentalissimi illius quarentene quorum decima hospitali		*ij acr'* Joh.

4 ♘ 26   Quarentena iacens transversa ad capud boriale.iiij.selionum hospitalis
predicti ultimo dictorum et abuttat ad capud sum occidentale super
smalemade

mo		ij.seliones Stephani Moriz senioris quondam Johanis Marchall primi		
sine balca		et australissimi illius quarentene et unus selio est forera		*j acr* Rad.
		ij.seliones hospitalis predicti iuxta		*j acr* Egid.
†		iij.seliones de terra universitatis iuxta		*iij rod* Rotund.
gun(*vile*)		ij.seliones Thome de Audele iuxta *bokenham*		*di acr* Rotund.

ix.seliones cum sua forera Roberti de Brigham iuxta et sunt ultimi
et borialissimi illius quarentene et iuxta Wlwyes dich   *j acr j rod* Egid.

5 ♘ 27   Quarentena in the Heerne vocata le erbeer iacens transversa ad capud
orientale.x.selionum cum una gora Galfridi busshell qui iacent
iuxta.ix.seliones Roberti de brygham ultimo dictos mediante Wlwyes
dich antedicta et abuttant ad capud eorum occidentale super
Coteweye

ij.seliones cum sua forera ad capud eorum australe Galfridi busshell *abuttant ad*
quondam Simonis de Refham primi et occidentalissimi illius   *capud australe*
quarentene quorum primus selio est forera et abuttat super   *super wlwyes dych*
Wlwyes Dich   *iij rod* Egid.

xiij.seliones cum sua forera ad capud eorum australe Rogeri de
harleston   *iij acr*

3b

[fol. 15v.]

quondam Willelmi Ward iuxta et iacent iuxta Wlwyes Dich predictam
et abuttant ad capud eorum australe super dictam Wlwyes dich et
iacent ibidem in angulo   Egid.

6 ♘ 28   Quarentena iacens transversa ad capud boriale quarentene ultimo
dicte et abuttat ad capud suum orientale super Wlwyes dich
predictam // Seliones illius quarentene debent computari ad capud
eorum orientale

MIDDELFELD

mo	j.selio Stephani Moriz quondam Johanis de Comberton primus et australissimus illius quarentene et est in parte forera ad capud suum orientale	*di acr'* Rad.
Inter.ij. balcas	⎰ j.selio Prioris de huntyngdon iuxta	*j rod* Rad.
b	⎱ j.selio monialum de bech iuxta et est in parte forera ad capud suum orientale	*di acr'* Par.
	viij.seliones Johanis de Cotton quondam Thome de Comberton iuxta et sunt curtiores aliis selionibus antedictis ad capud eorum orientale	*ij acr* Rad.
	j.selio Ricardi Tuliet iuxta *iij acr* [deleted]	*iij rod* Rad.
	ij.seliones Prioris de huntyngdon iuxta et sunt ultimi et borialissimi illius quarentene et ultimus selio est forera	*iij rod* Rad.
b	ij.seliones monialum de bech iacentes transversi ad capud orientale.viij.selionum Johanis de Cotton ultimo dictorum et iacent iuxta Wlwyes dich predictam quorum occidentalis selio est in parte forera ad capud suum australe et longior alio selione ad capud suum boriale et ambo sunt gores ad capud ultimo dictum	*di acr* Par.
7 ♮ 29	Quarentena abuttans ad capud suum Australe super foreram Prioris de Huntyngdon ultimo dictam / / Seliones illius quarentene debent computari ad capud eorum australe	
sine balca	⎧ iiij.seliones Nicholai crochman primi et occidentalissimi illius ⎪ quarentene et primus selio est forera	*ij acr* Rad.
	⎨ ij.seliones Willelmi de lolleworth alutarii quondam Johanis de ⎪ barton iuxta et solebant esse.iij.	*j acr* Par.
	⎩ ij.seliones Johanis Sherwynd quondam Willelmi Tuliet iuxta et solebant esse.iij.	*iij rod* Rad.

[fol. 16r.]   16

c cχ	⎰ j.selio Collegii predicti quondam Galfridi Seman iuxta	*di acr' b* Egid.
Inter.ij. balcas	⎱ j.selio monialum de bech iuxta	*di acr* Par.
.*b*.		
	j.selio hospitalis predicti iuxta	*di acr* Egid.
mo	j.selio Stephani Moriz senioris quondam Redhod iuxta	*di acr* Rad.
c cχ	ij.seliones Collegii predicti quondam Thome de Cantebrigia iuxta	*l acr b* Rad.
.*b*.	v.seliones monialum de bech qui vocatur porthors dole iuxta et iacent in longitudine in parte versus austrum iuxta predictos.ij. seliones dictarum monialum de beche qui ambo sunt gores ut supradictum est et in parte versus boriam iuxta Wlwyes dich predictam et sunt ultimi et orientalissimi illius quarentene	*ii acr* Par.
8 ♮ 30	Quarentene iacens transversa ad capud boriale quarentene ultimo dicte / / Seliones illius quarentene debent computari ad capud eorum orientale	
	xij.seliones hospitalis predicti qui vocantur brunneforthedole et abuttant ad capud eorum orientale super Wlwyes dich predictam quorum primus selio est longior aliis selionibus sequentibus ad capud suum occidentale et est forera et sunt primi et australissimi illius quarentene	*vj acr* Egid.

THE WEST FIELDS OF CAMBRIDGE

	vj.seliones parochialis ecclesie sancti sepulcri quondam Roberti tuliet iuxta et sunt longiores aliis selionibus antedictis ad capud eorum orientale et abuttant ut supra	*iij acr* Rad.
	ij.seliones Rogeri de harleston quondam Galfridi Seman iuxta et abuttant ut supra	*j acr* Par.
	j.selio hospitalis predicti iuxta et abuttat ut supra	*j rod* Egid.
	j.selio cantarie beate marie ecclesie clementis iuxta et abuttat ut supra	*j acr* Egid.
.b.	j.selio monialum de bech iuxta et abuttat super le brook	*di acr* Par.
	j.selio Rogeri de harleston quondam Willelmi Ward iuxta abuttans super le brook	*di acr* Botulph *nota*
	j.selio hospitalis predicti iuxta et abuttat super le brook	*ij rod* Egid.
*c*χ *c*	j.selio Collegii predicti quondam Thome de Cantebrigia iuxta abuttans super viam Sancti Neoti	*di acr b g* Rad.
	j.selio Rogeri de harleston quondam Galfridi Seman iuxta et est gorred ad capud suum occidentale et abuttat super viam predictam	*di acr* Egid.

4b

[fol. 16v.]

et dicti vij. seliones iacent inter.ij. balcas *c*χ *c*	iij.seliones qui solebant esse.v.dicti Rogeri quondam Roberti Goldsmith iuxta quorum primus selio extendit se in longitudine per.ij.quarentenas et tertius selio est curtior secundo ad capud suum occidentale et abuttat super viam predictam	*v rod* Par.

	j.selio Collegii predicti quondam Thome de Cantebrigia iuxta et est longior selione proxima precedente ac etiam aliis selionibus sequentibus ad capud suam occidentale et abuttat super viam predictam	*b* *iij rod* Rad.
.b. Inter.ij. balcas boyle *c*χ *c*	⎧ j.selio monialum de bech iuxta et abuttat super viam predictam ⎨ iiij.seliones qui solebant esse.viij.Rogeri de harleston quondam ⎩ Roberti Goldsmith iuxta et abuttant super viam predictam	*di acr* Par. *iij acr* Par.
	ij.seliones Collegii predicti quondam Thome de Cantebrigia iuxta et abuttant super viam predictam	*1 acr et di* Botulph
	j.selio Prioris de Bernewell iuxta et abuttat super viam predictam	*di acr* Egid.
nota	j.selio Willelmi Aleyn de Bokesworth iuxta et abuttat super viam predictam	(∴ *di acr* Botulph
	v.seliones Prioris de Bernewell iuxta et abuttant super viam predictam	*ij acr di* Egid.
*c*χ	ij.seliones Collegii predicti quondam Thome de Cantebrigia iuxta et abuttant super viam predictam	*j acr* Rad.
	vj.seliones qui solebant esse.vij.Prioris de Bernewell iuxta abuttantes super viam predictam	*iij acr* Egid.
*c*χ *di a*	j.selio Collegii predicti quondam Thome de Cantebrigia iuxta et abuttat super viam predictam	Rad.
	iiij.seliones Rogeri de Harleston quondam Roberti Goldsmith iuxta et abuttant super dictam viam	*ij acr* Par.
	j.selio dicti Rogeri quondam Roberti longe iuxta et abuttat super dictam viam	*di acr* Par.

110

MIDDELFELD

	iiij.seliones qui solebant esse.vj.dicti Rogeri quondam Roberti Goldsmith iuxta et abuttant super viam predictam	*iij acr* Par.
cχ *l acr et di*	iiij.seliones Collegii predicti quondam Galfridi Seman iuxta et abuttant super viam predictam	Egid.
	ij.seliones Rogeri de harleston quondam Willelmi de lolleworth iuxta abuttantes super dictam viam	*l acr* Par.
	ij.seliones hospitalis predicti iuxta et abuttant super viam predictam	*l acr* Egid.
.b.	j.selio monialum de bech iuxta et abuttat super viam predictam	*di acr* Par.
cχ *ij rod*	⎧ ij.seliones Collegii predicti quondam Galfridi Seman iuxta et abuttant super viam predictam	Egid.
nter.ij. balcas ⎨	j.selio monialum sancte Radegunde iuxta et abuttat super viam predictam	*l rod di* Par.
clar' hall	⎩ j.selio Thome marbilthorp iuxta abuttans super viam predictam	*iij rod* Par.
	vij.seliones Prioris de Bernewell iuxta abuttantes super viam predictam	*iij acr' di acr* Egid
cχ	ij.seliones Collegii predicti quondam Galfridi Seman iuxta abuttantes super dictam viam	[*j acr*] Par.
nter.ij. balcas	⎰ j.selio hospitalis predicti iuxta abuttans super dictam viam	*di acr* Egid.
	⎱ j.selio Prioris de Bernewell iuxta abuttans super dictam viam	*di acr* Egid.

[fol. 17r.]   17

u	*cχ*	j.selio Collegii predicti quondam Thome de Cantebrigia iuxta et abuttat super dictam viam (*di* deleted) .*j. acr*	Par. secundum usum
riffin<	*mo*	ij.seliones Stephani Moriz quondam patris sui iuxta et abuttant super dictam viam	*j acr* Egid.
	mo	ij.seliones dicti Stephani quondam Willelmi de bekeswell iuxta abuttantes super dictam viam	*j acr* Par.
		iiij.seliones hospitalis predicti iuxta et abuttant super dictam viam	*ij acr di* Rad.
p	*cχ*	ij.seliones Collegii predicti quondam Galfridi Seman iuxta et abuttant super dictam viam	*j acr* Par. ca
	@	j.selio Rogeri de harleston quondam Willelmi de lolleworth iuxta abuttans super dictam viam	*di acr* Par.
		j.selio Prioris de bernewell iuxta et abuttat super dictam viam	*di acr* Egid.
		v.seloines hospitalis predicti iuxta abuttantes super dictam viam	*ij acr* Egid.
		ij.seliones Rogeri de harleston quondam Willelmi de lolleworth iuxta et sunt longiores aliis selionibus antedictis ad capud eorum occidentale abuttantes super dictam viam	*di acr* Par.
cχ		j.selio Prioris de huntyngdon iuxta et est curtior aliis selionibus antedictis ad capud suum orientale abuttans super foreram Collegii predicti quondam Galfridi Seman ad dictum capud	*di acr* Rad.
	mo	ij.seliones Stephani Moriz junioris quondam Johanis Pittok iuxta abuttantes super dictam foreram	*j acr'* Par.
nter.ij. balcas	⎰	ij.seliones Roberti longe iuxta et abuttant super foreram predictam	*j acr'* Par.
	⎱	j.selio Thome atcherch de howes quondam Johanis blancpayn iuxta abuttans super dictam foreram	*di acr'* Rad.

111

THE WEST FIELDS OF CAMBRIDGE

f ♭ cχ		⎧j.selio Collegii quondam Thome de Cantebrigia iuxta abuttans super dictam foreram	di acr' Rad. . ⋅ . di acr
cχ vj acr		⎨xj.seliones dicti Collegii quondam Galfridi Seman iuxta abuttantes super dictam foreram quorum.iiij.ultimi seliones sunt longiores aliis selionibus antedictis ad capud eorum occidentale	vj. acr Egid.
		iiij.seliones Prioris de Bernewell iuxta et abuttant super foreram predictam quorum ultimus selio est forera et sunt ultimi et borialissimi illius quarentene	ij acr Egid.
9 ♭ 31		Quarentena ultra altam crucem iacens transversa ad campum de cotes / / Omnes seliones illius quarentene abuttant ad capud eorum orientale super viam sancti Neoti antedictam exceptis.vj.selionibus qui sunt primi et australissimi illius quarentene et curtiores aliis selionibus sequentibus et abuttant super foreram Prioris de bernewell ad dictum capud *et Incipe in parte australi*	8
	mo	j.selio Thome Moriz quondam domini Johanis Saverey de Cotes capellani et australissimus illius quarentene et est forera	iij rod Rad 5b

[fol. 17v.]

co cχ		j.selio Collegii predicti quondam Galfridi Seman iuxta	di acr' Egid
	@	ij.seliones Rogeri de harleston quondam Willelmi de lolleworth iuxta	j acr Par.
		ij.seliones Prioris de bernewell iuxta	j acr Egid
cχ ij Elies acr et di		⎧ij.seliones Collegii predicti quondam Galfridi Seman iuxta et sunt longiores aliis selionibus antedictis ad capud eorum orientale et primus selio est in parte forera ad capud predictum	Egid.
co cχ ij acr		⎩iij.seliones dicti Collegii quondam dicti Galfridi et quondam Johanis atte Gren iuxta	Rad
co v ij acr		ix.seliones eiusdem Collegii quondam dicti Galfridi iuxta	v acr Egid
		j.selio hospitalis predicti iuxta	j acr Rad
sine balca		⎧ij.seliones Prioris de huntyngdon iuxta	j acr Rad
		⎩iij.seliones Collegii predicti quondam Galfridi Seman quorum ultimus selio est quasi forera ad campum de Cotes et sunt ultimi et borialissimi illius quarentene	j acr et rod Egid
sheriffes dole 10 ♭ 32		Quarentena iacens in longitudine ad capud occidentale quarentene ultimo dicte *quarentena prima ex latere boriali quarentene*	
sine balca		⎧j.selio Prioris de Bernewell primus et borialissimus illius quarentene et iuxta moram et campum de Cotes quadam lata balca mediante et est longior aliis selionibus sequentibus ad capud suum orientale	di acr Egid
	mo	⎩j.selio Stephani Moriz senioris quondam Johanis Redhod iuxta	di acr Egid
		j.selio Prioris de bernewell iuxta	di acr Egid
		j.selio hospitalis predicti iuxta	di acr Egid
mo		vj.seliones Thome Moriz iuxta	iij acr Egid
modo sunt x		xj.seliones Prioris de huntyngdon iuxta nunc ix	iiij acr Rad
		ij.seliones qui solebant esse.iiij.Thome de Audele iuxta	j acr di Par
cχ		ij.seliones Collegii predicti quondam Thome de Cantebrigia iuxta	j acr Rad
		ij.seliones Rogeri de Baldeston de Cotes quondam Johanis de Comberton iuxta	j acr Rad

MIDDELFELD

		iij.seliones Prioris de Bernewell iuxta	*j acr* Egid.
mo		iij.seliones Thome Moriz iuxta et ultimus selio est forera et sunt ultimi et australissimi illius quarentene	*j acr' di* Egid.

[fol. 18r.] 18

11 ♘ 33   Quarentena iacens transversa ad capud orientale.iij.selionum Thome
Moriz ultimo dictorum / / Seliones illius quarentene debent
computari ad capud eorum boriale

	xiij.seliones Prioris de Bernewell primi et occidentalissimi illius quarentene quorum.iiij.primi seliones extendunt se per.ij.quarentenas et primus selio est in parte forera	*iij acr'* Egid.
Ely *cχ*	v.seliones et.j.gore valde lati cum.j.lata butta iacentes ex parte occidentale ad capud eorum australe Collegii predicti quondam Galfridi Seman iuxta et sunt longiores aliis selionibus antedictis ad capud predictum	(: *iij acr* Egid.
sine balca	⎧ iij.seliones Prioris de Bernewell iuxta et sunt longiores aliis selionibus ⎨ antedictis ad capud eorum boriale et primus selio est in parte forera ⎩ ad capud predictum	*j acr' di* Egid.
	j.selio Rogeri de harleston quondam Willelmi de lolleworth iuxta et est ultimus et orientalissimus illius quarentene et iuxta viam sancti neoti antedictam	*di acr* Par.

12 ♘ 34   Quarentena curta continens.x.buttes Collegii predicti quondam
f cχ   Galfridi Seman iacens transversa ad capud australe.iij.selionum
Prioris de Bernewell ultimo dictorum et abuttat ad capud suum
orientale super viam Neoti iacent' iuxta altam crucem   *iij acr* [sic]

f cχ	⎧ xj.buttes Collegii predicti quondam Galfridi Seman primi et borialissimi ⎨ illius quarentene quarum prima butta est forera	Egid.
f cχ	⎨ viij.buttes eiusdem Collegii quondam dicti Galfridi iuxta quarum ⎩ ultima butta est forera	Rad.

id summa iij acr

13 ♘ 35   Quarentena iacens iuxta partem occidentalem vie sancti Neoti
alta ✠   predicte iuxta altam crucem antedictam / / Seliones illius quarentene
debent computari ad capud eorum australe

| *f cχ* | iiij.longi seliones et.iiij.buttes Collegii predicti quondam Galfridi Seman primi et orientalissimi illius quarentene et iuxta viam sancti Neoti predictam et abuttant ad capud eorum boriale super buttam dicti Collegii quondam dicti Galfridi ultimo dictam que est | *iij acr et di* |

6b

[fol. 18v.]

	forera ut supradictum est quorum quartus selio est in parte forera ad capud australe		Egid.
mo	viij. longi seliones Thome Moriz iuxta quorum primus selio est in parte forera videlicet forera in medio	*nunc vj*	*iiij acr* Rotund'

113

THE WEST FIELDS OF CAMBRIDGE

	iiij.seliones Prioris de bernewell extendentes se in longitudine per.ij. quarentenas ut supradictum est in principio secunde quarentene precedente iuxta	*ij acr di* Egid.
co cχ	j.selio Collegio predicti quondam Galfridi Seman iuxta	*di acr* Egid.
mo	ij.seliones Stephani Moriz junioris quondam Johanis Pittok iuxta	*v rod* Par.
	ij.seliones Prioris de bernewell iuxta	*1 acr 1 rod* Egid.
Elys cχ	ij.seliones Collegii predicti quondam Galfridi Seman iuxta	*iij rod* Egid.
mo	ij.seliones Stephani Moriz senioris quondam patris sui	*j acr rod* Egid.
	j.selio qui solebat esse.ij.sancti Egidii iuxta	*j acr* Par.
	j.selio Prioris de Bernewell iuxta	*di acr* Egid.
co cχ nunc 8	ix.seliones Collegii predicti quondam Galfridi Seman iuxta quorum.ij. primi seliones sunt longiores aliis selionibus sequentibus ad capud eorum australe et secundus selio est in parte forera ad capud predictum	*iij acr et di* Egid.
.v.acr	viij.seliones monialum sancte Radegunde iuxta quorum quartus et quintus seliones sunt in parte forera ad capud eorum australe et.iiij. ultimi seliones sunt curtiores. iiij.selionibus precedentibus ad utrumque capud et sunt ultimi et occidentalissimi illius quarentene et ultimus selio est forera	*iij acr* Par.
14 ♭ 36	Quarentena abuttans ad capud suum occidentale super Endelesweye // Seliones illius quarentene debent computari ad capud eorum occidentale	
cχ	⎧ xj.seliones Collegii predicti quondam Thome de Cantebrigia primi et borialissimi illius quarentene quorum.vj.primi seliones *modo. v.* abuttant ad capud eorum occidentale super campum de Cotes	
sine balca	⎨ et.vj. ultimi seliones abuttant super endelesweye ad capud predictum et primus selio dictorum.xj.selionum est in parte forera ad campum antedictum	*iij acr et di* Rad.
	⎩ vj.seliones cum sua forera Prioris de Bernewell iuxta	*iij acr* Egid.
	m. dyer occupat terras collegii in tenura prioris de B.	

[fol. 19r.]   19

sine balca	ij.seliones Johanis de Cotton quondam Thome de Comberton et quondam Johanis Tableter iuxta	*j acr* Rad.
Inter foreram et balcam *f cχ*	⎧ xiiij.seliones cum sua forera cum.j.gora ex parte australe ad capud eorum orientale Prioris de Bernewell iuxta quadam parva fovea mediante quorum primus secundus et tertius seliones sunt in parte forera ⎨ viij.seliones Collegii predicti quondam Thome de Cantebrigia iuxta et sunt longiores aliis selionibus antedictis ad capud eorum orientale	*vij acr dukdole* Rad. *ij acr et di* Rad.
	⎩ ij.seliones Prioris de Bernewell iuxta	*j acr* Egid.
mo	ij.seliones Stephani Moriz junioris quondam Johanis Pittok iuxta	*j acr* Par.
mo	v.seliones Thome Moriz iuxta	*ij acr et di* Egid.
	iij.seliones monialum sancte Radegunde iuxta et sunt longiores aliis selionibus antedictis in duplum	*vj rod* Egid.
	iij.seliones hospitalis predicti iuxta	*vj rod* Egid.
mo Inter.ij. balcas	⎧ i.selio Stephani Moriz junioris quondam Johanis Pittok iuxta ⎨ ij.seliones Rogeri de harleston quondam Willelmi de lolleworth ⎩ iuxta	*j rod* Par. *j acr* Par.

MIDDELFELD

	xij.seliones monialum sancte Radegunde iuxta *nunc xj*	*iiij acr* Par.
	iij.seliones Roberti Tuliet iuxta *nunc ij*	*j acr* Egid.
lu – *cχ*	ij.seliones Collegii predicti quondam Thome de Cantebrigia iuxta	*j acr et di* Rad.
/ *cχ*	ij.seliones eiusdem Collegii quondam Johanis Redhod iuxta	Rad.
.*b*.	vj.seliones monialum de bech iuxta	*iiij acr* Par.
	xij.seliones monialum sancte Radegunde iuxta	*vj acr* Par.
	j.selio Rogeri de harleston quondam Willelmi de lolleworth iuxta	*1 rod di* Par.
mo	j.selio Stephani Moriz senioris quondam Johanis Redhod iuxta	*1 rod di* Par.
lu cχ	j.selio Collegii predicti quondam Thome de Cantebrigia iuxta	*1 rod* Rad.
.*b*.	ij.seliones monialum de bech iuxta	*di acr* Par.
	iiij.seliones qui solebant esse.v.Prioris de bernewell iuxta	*1 acr di* Egid.
	j.selio Ricardi Tuliet iuxta *clar' hall*	*di acr* Rad.
.*b*.	⎧ j.selio qui solebat esse.ij.monialum de bech iuxta	*1 acr* Par.
Inter.ij. balcas	⎨ j.selio Collegii predicti quondam Galfridi Seman iuxta	*di acr* Egid.
lu cχ	⎩ iij.seliones Roberti Tuliet iuxta j. acr	Rad.
mo	ij.seliones Stephani Moriz senioris quondam Willelmi de lolleworth qui vocatur	

.*f*

[fol. 19v.]

	le gildenaker iuxta	*j acr* Egid.
	ij.seliones qui solebant esse.iij.Prioris de bernewell iuxta	*j acr di* Rotund'
	iij.seliones hospitalis predicti iuxta	*j acr di* Egid.
mo	⎧ j.selio Stephani Moriz senioris quondam Willelmi de bokeswell iuxta	*j rod* Par.
Inter.ij. balcas	⎨ Omnes seliones predicti abuttant ad capud eorum occidentale super endlesweye exceptis.vj.selionibus qui sunt primi et borialissimi illius quarentene abuttantibus super campum de cotes ut supradictum est	
	⎨ ij.seliones Rogeri de Harleston quondam Willelmi Ward iuxta et sunt longiores aliis selionibus antedictis ad capud eorum occidentale et extendunt se in longitudine ultra Endlesweye Path	*j acr* Egid.
cχ per	j.selio qui solebat esse.ij.Prioris de bernewell iuxta	*j acr* Egid.
excambium	iij.seliones Rogeri de Harleston quondam Willelmi de lolleworth	*j acr* Par.
prope mor' bernys	iuxta	
mo	v.seliones Stephani Moriz senioris quondam Willelmi de Bekeswell iuxta	*ij acr di* Par.
	⎧ ij.seliones Prioris de Bernewell iuxta et sunt curtiores aliis selionibus antedictis ad capud eorum orientale	*di acr* Egid.
sine balca	⎨ ij.seliones hospitalis predicti iuxta quorum primus selio est longior secundo ac etiam aliis.ij.selionibus precedentibus ad capud suum orientale	*ij rod* Egid.
	⎨ iiij.seliones Rogeri de harleston quondam Willelmi de Lolleworth iuxta quorum primus selio extendit se in longitudine per.ij.quarentenas videlicet usque ad viam sancti neoti antedictam	*v rod* Par.
	⎩ j.selio dicti Rogeri quondam Johanis de berton iuxta et est ultimus et australissimus illius quarentene et iuxta.vj.seliones dicti Rogeri quondam dicti Johanis de berton extendentes se in longitudine usque ad Coteweye	*j rod* Egid.

THE WEST FIELDS OF CAMBRIDGE

15 ♭ *37*		Quarentena abuttans ad capud suum occidentale super Coteweye predictam et vocatur Thorpis croft furlong	
		x.seliones et.j.gora cum sua forera ad capud eorum occidentale Galfridi busshell quondam Simonis de refham primi et australissimi	
		illius quarentene et iuxta Wlwyes dich	*iij acr* Egid.
		j.selio hospitalis predicti iuxta	*di acr* Egid.
	mo	j.selio Stephani Moriz senioris quondam patris sui iuxta	*di acr* Egid.

[fol. 20r.] 20

	mo	iij.seliones Thome Moriz iuxta *lovel*	*j acr* Egid.
gun(*vile*)		⎧ ij.seliones Thome de Audele iuxta *bokenham quondam rawlyns*	*iij rod di* Par.
Inter.ij. balcas		⎨ j.selio Johanis Pilet quondam patris sui iuxta	*j rod* Par.
	b	⎩ j.selio monialum de bech iuxta	*j rod di* Par.
Cleg		iij.seliones Ricardi martyn quondam Johanis martyn iuxta et vocatur le gildenaker	*j acr* Rad.
c cχ		iiij.seliones Collegii predicti quondam Galfridi Seman iuxta et sunt curtiores aliis selionibus antedictis ad capud eorum orientale *j acr et iij rod*	*b* Par.
		xj.seliones Prioris de Bernewell iuxta quorum septimus selio est in parte forera ad capud suum orientale et.iiij.ultimi seliones sunt curtiores aliis selionibus antedictis ad capud predictum	*iij acr* Rotund.
	o	iiij.seliones ocupati per predictum Priorem de terra hospitalis predicti prout patet per rotulos terrarum eiusdem hospitalis iuxta	*v rod* Rotund.
	o	j.selio ocupatus per predictum Priorem de terra eiusdem hospitalis prout patet per dictos rotulos et per libros de decimis de bernewell iuxta	*j rod* Egid.
		Et predicti.xx.seliones iacent inter.ij.balcas	
c cχ		ij.seliones Collegii predicti quondam Galfridi Seman iuxta *j acr*	*b* Egid.
		ij.seliones hospitalis predicti iuxta	*iij rod* Egid.
c cχ iamj		ij.seliones Collegii predicti quondam Thome de Cantebrigia iuxta *iij rod*	*b* Rad.
c cχ ij acr		vj.seliones Prioris de Bernewell iuxta *nunc Corporis Christi b* (∴ *excambium j acr di* Egid. *g*	
		ij.seliones Johanis Pilet quondam patris sui iuxta	*v rod* Par
		ij.seliones hospitalis predicti iuxta quorum secundus selio est longior aliis selionibus antedictis ad capud suum orientale quorum decima	*j acr* Hospitali dicto
		j.selio cantarie beate marie ecclesie sancti clementis iuxta	*iij rod* Rotund.
		v.seliones Thome Moriz iuxta	*ij acr* Egid.
		iij.seliones Cantarie beate marie ecclesie sancti sepulcri iuxta	*j acr di* Egid.
c cχ		j.selio Collegii predicti quondam Galfridi Seman iuxta	*di acr b* Egid.
	mo	iij.seliones Thome Moriz iuxta	*j acr* Egid.
		vj.seliones Prioris de Bernewell iuxta	*iij acr* Egid.
		vj.seliones Rogeri de Harleston quondam Johanis de Berton iuxta	*iij acr* Egid.
c cχ		j.selio Collegii predicti quondam Thome de Cantebrigia iuxta et est curtior aliis selionibus antedictis ad capud suum orientale *di acr*	(∴ *b* Rad.

[fol. 20v.]

		j.selio Prioris de Huntyngdon iuxta	*di acr* Rad.
griffyn mo		ij.seliones Stephani Moriz senioris quondam Simonis Howdelowe iuxta	*j acr* Par.

MIDDELFELD

Inter.ij. balcas	⎧ iij.seliones Rogeri de Harleston quondam Roberti Goldsmith iuxta	j acr Par.
mo	⎨ j.selio Nicholai Crocheman iuxta	j rod Par.
ca cχ	⎩ ij.seliones Thome Moriz iuxta	di acr Par.
	iij.seliones Collegii predicti quondam Galfridi Seman iuxta	j acr et di rod Par.
	iiij.seliones qui solebant esse.v.hospitalis predicti iuxta	ij acr Par.
mo	iiij.seliones Thome Moriz iuxta	j acr di Par.
	j.selio Rogeri de harleston quondam Roberti Goldsmith iuxta	j rod Par.
	j.selio hospitalis predicti iuxta	j rod Par.
o	ij.seliones monialum sancte Radegunde iuxta quorum ultimus selio est ocupatus per Cantariam beate marie ecclesie sancti sepulcri	di acr Par.
	⎧ j.selio Roberti Longe iuxta	di acr Par.
†	⎪ j.selio ocupatus per Rogerum de harleston	di acr Egid.
Inter.ij. balcas	⎨ iiij.seliones dicti Rogeri quondam Galfridi Seman iuxta	vj rod Par.
	⎪ iiij.seliones dicti Rogeri quondam dicti Galfridi iuxta	j acr di Egid.
	⎩ iiiij.seliones dicti Rogeri quondam Roberti Goldsmith iuxta	ij acr Par.
c cχ	vj.seliones Collegii predicti quondam Thome de Cantebrigia iuxta ij acr	b Rad.
	iiij.seliones Cantarie beate marie ecclesie sancti sepulcri iuxta	ij acr Egid.
	⎧ ij.seliones hospitalis predicti iuxta	di acr Egid.
mo	⎨ j.selio Stephani Moriz junioris quondam hugonis Pittok iuxta	j rod Egid.
Inter.ij. balcas	⎩ j.selio Cantarie beate marie ecclesie sancti sepulcri iuxta	j rod Egid.
mo	iiij.seliones Thome Moriz iuxta	ij acr Egid.
	iij.seliones Cantarie beate marie ecclesie sancti sepulcri iuxta	j acr di Egid.
	ij.seliones Roberti Longe quondam Johanis Tableter iuxta	iij rod Egid.
ca cχ	j.selio Collegii predicti quondam Thome de Cantebrigia iuxta	iij rod Rad.
mo	j.selio Stephani Moriz senioris quondam Johanis Redhod iuxta	di acr Rad.
b	v.seliones monialum de bech iuxta	ij acr di Par.
mo	iij.seliones Stephani Moriz junioris quondam Johanis Pittok iuxta	j acr di Egid.

[fol. 21r.] 21

inter duas balcas	⎧ ij.seliones hospitalis predicti iuxta	j acr Egid.
	⎩ j.selio Stephani Moriz junioris quondam Johanis Pittok iuxta	j acr Par.
mo	ij.seliones Cantarie beate marie ecclesie sancti sepulcri iuxta	j acr Egid.
	ij.seliones hospitalis predicti iuxta	j acr Rad.
inter duas balcas	⎧ ij.seliones Johanis de Cotton quondam Thome de Comberton iuxta	j acr Rad.
	⎩ ij.seliones Roberti Tuliet iuxta j acra clar' hal	j acr Rotund.
	ij.seliones cum.j.gora ex parte australe ad capud eorum orientale	
mo	Stephani Moriz senioris quondam patris sui qui vocatur le gored aker iuxta	j acr Egid.
	ij.seliones Prioris de Bernewell iuxta	j acr Egid.
	j.selio Stephani Moriz senioris quondam Johanis Redhod iuxta thorpescroft Hic incipit thorpescroft	di acr Par.
△	j.selio Roberti longe quondam Johanis Tableter iuxta	di acr Par.
mo	v.seliones Stephani Moriz senioris quondam Johanis Redhod iuxta	ij acr Par.

THE WEST FIELDS OF CAMBRIDGE

	mo	v.seliones Stephani Moriz junioris iuxta	*ij acr* Par.
		iij.seliones Roberti Longe iuxta *et a gore*	*j acr di* Par.
		j.selio cum sua forera cantarie beate marie ecclesie de Cotes iuxta	*j rod* Par.
Inter		j.selio cum sua forera Willelmi de Thaxsted iuxta	*j rod* Par.
ij	*mo*	ij.seliones cum sua forera Stephani Moriz junioris quondam Johanis Pittokes iuxta	*j rod* (deleted) *j acr* Par.
balcas		vj.seliones cum sua forera Stephani Moriz senioris iuxta	*ij acr di* Par.
		ij.seliones cum sua forera Johanis Templeman de Cotes iuxta quorum primus selio quondam Roberti Longe	*di acr di rod* Par
		ij.seliones cum sua forera quondam Roberti Seman nunc in manibus Prioris de Bernewell iuxta	*j acr di* Par.
		iij.seliones cum sua forera Johanis de Coton quondam Thome de de Comberton iuxta	*j acr di* Par.
		iij.seliones cum sua forera Roberti Longe quondam Johanis Tableter iuxta	*j acr di* Par.
		iij.seliones cum sua forera Nicholai Wymak de Cotes quondam Willelmi de Bekeswell iuxta	*j acr di* Par.
		ij.seliones cum sua forera Johanis Wymak de Cotes quondam Roberti Longe iuxta	*iij rod* Par.
sine balca		ij.seliones cum sua forera Cantarie ecclesie clementis quondam Ricardi Youn iuxta	*di acr* Par.
		ij.seliones cum suis foreris Johanis Wymark de Cotes quondam Roberti longe iuxta quorum ultimus selio est quasi forera ad campum de Cotes et sunt ultimi et borialissimi illius quarentene	*iij rod* Par.

Thorpescroft ys a grove of iij or iiij selyons brode at the weste ende by coton weye on the este parte of ytt ad tale signum

[fol. 21v.]   Cotton

16 ♭ 31   Quarentena ex altera parte de Coteweye predicta vocatur Alderman hill et abuttat ad capud suum orientale super Coteweye predictam

	j.lata gora et.iij.seliones cum sua forera Johanis Templeman de Cotes qui vocantur Sparwes Croft primi et borialissimi illius quarentene et iuxta campum de Cotes mediante de Daleweye	*j acr* Egid.
mo	iij.seliones Stephani Moriz senioris iuxta	*j acr* Rad.
f cχ	j.selio Collegii predicti quondam Johanis Redhod iuxta	*j rod* Rad.
	ij.seliones Roberti Longe quondam Johanis Tableter iuxta *clar' hall j acr* Par.
	ij.seliones Cantarie beate marie ecclesie sancti sepulcri iuxta	*j acr* Rad.
Katerine hall	ij.seliones Roberti Longe quondam Johanis Tableter iuxta *Katerine j acr* Egid.
	j.selio Cantarie beate marie ecclesie de cotez iuxta	*di acr* Rad.
mo	iij.seliones et.j.gora Stephani Moriz junioris quondam Johanis Pittok iuxta et vocantur le Goredaker	*j acr* Egid.
nota	j.selio Prioris de Bernewell iuxta	*di acr* Botul'.
Ely *cχ*	j.selio Collegii predicti quondam Thome de Cantebrigia iuxta	*di acra* Rad.
	viij.seliones cum sua forera Rogeri harleston quondam Ricardi Tableter iuxta	*iij acr* Par.
	vij.seliones cum sua forera dicti Rogeri quondam Willelmi Ward iuxta	*ij acr di* Egid.
mo	ix.seliones Thome Moriz iuxta	*iiij acr di* Egid.
	vj.seliones Prioris de Bernewell iuxta	*iij acr* Egid.

MIDDELFELD

	mo	v.seliones Thome Moriz iuxta *nunc iiij*	*ij acr* Egid.
ca cχ		j.selio ocupatus per Collegium predictum tamquam terram Galfridi quondam Roberti in ye lane ut patet per libros de decimis de bernewell iuxta	*di acr* Egid.
	mo	iij.seliones Thome Moriz iuxta	*v rod.* Egid.
cχ		iiij.seliones Collegii predicti quondam Galfridi Seman iuxta	*ij acr b* Egid.
nota		j.selio Prioris de Bernewll iuxta	*di acr* Botul'
ca cχ		j.selio Collegii predicti quondam Galfridi Seman iuxta	*di acr* Par.
	mo	v.seliones Thome Moriz iuxta	*ij acr di* Egid.
		j.selio Cantarie beate marie ecclesie sancti sepulcri iuxta	*di acr* Egid.

[fol. 22r.] 22

	mo	⎧ j.selio Thome Moriz iuxta	*di acr* Egid.
inter.ij. balcas		⎨ iiij.seliones Stephani Moriz senioris quondam patris sui	*ij acr* Par.
	mo	⎩ iuxta	
	mer'	⎧ j.selio de terra clericorum de merton iuxta	*j rod* Egid.
inter.ij. balcas		⎩ j.selio Collegii predicti quondam Thome de Cantebrigia iuxta	*j rod* Rad.*co* ⎫
cχ			*sa* ⎬
	mo	iiij.seliones Stephani Moriz senioris quondam patris sui iuxta	*ca* ⎭ *ij acr* Par.
		j.selio monialum sancte Radegunde iuxta	*iij rod* Par.
	mo	⎧ j.selio Stephani Moriz senioris quondam patris sui iuxta	*di acr* Egid.
inter.ij. balcas		⎨ j.selio Rogeri harleston quondam Johanis de Berton iuxta	*di acr* Rad.
		⎩ x.seliones dicti Rogeri quondam dicti Johanis de Berton iuxta	*v acr* Egid.
		ij.seliones cantarie beate marie ecclesie sancti sepulcri iuxta	*j acr* Egid.
		ij.seliones ocupati per Thomam marbylthorp iuxta et vocantur le blakaker.j.acra quondam Thome Blangronis *hospit.*	*j acr* Rad.
		xlv. seliones hospitalis predicti iuxta *nunc xlii*	*xxviij acr* Egid.
gun(*vile*)		xlj.seliones de terra mortimer iuxta *nunc 38*	*xx acr* Rad.
c cχ		vij.seliones Collegii predicti quondam Galfridi Seman iuxta	*j acr b* Egid. *3ac*
		j.selio hospitalis predicti iuxta	*di acr* Egid.
c cχ		ij.seliones Collegii predicti quondam Galfridi Seman et vocantur karlokaker	*ij acr di b* Botul' *nota*
cχ		iiij.seliones eiusdem Collegii quondam dicti Galfridi iuxta	Par.
†		ix.seliones de terra Universitatis iuxta tamen aliqui eorum sunt buttes et sunt ultimi et australissimi illius quarentene et iuxta bertonweye	*ij acr* Rotund'.

17 ▯ 39   Quarentena iacens in longitudine ad capud occidentale quarentene
ultimo dicte ex altera parte de **Aldermanhill** predicto / / Seliones
illius quarentene debent computari ad capud eorum occidentale

xxviij.seliones et.j.gora hospitalis predicti abuttantes ad capud eorum
occidentale super Ewyne Dych et sunt primi et australissimi   *xvj acr* Egid.
illius quarentene et iuxta bertonweye predictam *abuttant in parte*
in capite occidentali super Berton weye per omnes seliones et in parte
australi abbutant super lykylt grene et parte berton weye vertendo
directe versus berton.

THE WEST FIELDS OF CAMBRIDGE
[fol. 22v.]

	xiij.seliones cum sua forera Roberti Tuliet iuxta	*v acr cotton* Rad.
	xj.seliones Prioris de Bernewell iuxta	*iiij acr* Egid.
mo	⎧vj.seliones Thome Moriz iuxta	*ij acr* Rad.
inter.ij. balcas	⎨v.seliones monialum de bech quondam hugonis Pittok iuxta	*ij acr di* Par.
b		
	⎧j.selio Cantarie beate marie ecclesie clementis quondam Gilberti in	
inter.ij. balcas	⎨the lane iuxta	*di acr* Par.
	⎩ij.seliones hospitalis predicti iuxta	*j acr* Egid.
	ij.seliones Prioris de Bernewell iuxta	*iij rod* Egid.
	iij.seliones hospitalis predicti iuxta	*j acr di* Rad.
mo	⎧iiij.seliones Stephani Moriz senioris quondam patris sui iuxta	*ij acr* Egid.
inter.ij. balcas	⎩v.seliones Rogeri de harleston iuxta quorum.iiij.primi seliones quondam	
	Willelmi Ward et quintus selio quondam Ricardi Tableter iuxta	*ij acr di* Egid.
mo	⎧j.selio Stephani Moriz senioris quondam patris sui iuxta	*j rod di* Egid.
inter.ij. balc'	⎩j.selio Prioris de Bernewell iuxta	*j rod di* Egid.
	ij.seliones Rogeri de Harleston quondam Ricardi Tableter iuxta	*j acr* Par.
	⎧ij.seliones hospitalis predicti iuxta quorum secundus selio est in	
	⎪parte forera ad capud occidentale	*j acr* Egid.
inter.ij.	⎨v. seliones eiusdem hospitalis iuxta et sunt curtiores aliis selionibus	
balcas	⎪antedictis ad capud occidentale	*j acr j rod* Rotund
	⎩xj.seliones Prioris de Bernewell iuxta	*iiij acr* Rotund
cχ ca	ij.seliones Collegii predicti quondam Galfridi Seman iuxta quorum	
	secundus selio est in parte forera ad capud suum occidentale	*j acra* Par.
	ij.seliones Prioris de Bernewell iuxta et sunt curtiores aliis selionibus	
	antedictis ad capud eorum occidentale	*di acr* Egid.
mo	j.selio Stephani Moriz senioris quondam Simonis howdlo iuxta	*di acr* Rad.
	vj.seliones Prioris de Bernewell iuxta	*ij acr* Rad.
	vj.seliones hospitalis predicti iuxta	*ij acr* Egid.
cχ	iiij.seliones Collegii predicti quondam Galfridi Seman iuxta *v acr*	⎫ Par.
cχ ⎬*ca*	ix.seliones eiusdem Collegii quondam dicti Galfridi iuxta	⎭ Egid.
cχ ⎭	xiiij.seliones eiusdem Collegii quondam Thome de Cantebrigia iuxta	
		v acr Rad.
	⎧v.seliones Prioris de Bernewell iuxta	*ij acr* Egid.
inter.ij. balcas	⎩iiij.seliones hospitalis predicti iuxta	*ij acr* Egid.

[fol. 23r.] 23

mo	x.seliones Stephani Moriz senioris quondam patris sui iuxta	*v acr* Egid.
inter.ij. balcas	iij.seliones hospitalis predicti iuxta	*j acr* Rad.
	vij.seliones qui solebant esse.viij.cum sua forera Thome atte chirch	
	de howes quondam Willelmi lavenham iuxta et vocantur Sparwes	
	croft et sunt ultimi et borialissimi illius quarentene et iuxta campum	
	de cotes mediante le Dale Weye predicta	*iiij acr* Egid.

| 18 | ♭ | 40 | Quarentena iacens transversa ad capud occidentale.vij.selionum Thome at cherch ultimo dictorum et est.j.quarentena de le deddale *in the lowe* |

MIDDELFELD

sine balca	iiij.seliones Walteri Goldsmith quondam Willelmi atte cunditt	Cott'
	quorum primus selio est forera et sunt primi et orientalissimi illius quarentene *clar' hall*	*ij acr* Rad.
	v.seliones Willelmi Sherwynd quodam Willelmi Tuliet iuxta	*iij acr* Rad.
f cχ		
j acr	iij.seliones Collegii predicti quondam Johanis Redhod et quondam Alicie Fadersoul iuxta	Rad.

xv.seliones domini bartholomei burghassh militis qui vocantur
erlesdole iuxta mediante le Weie balk et sunt ultimi et   *erlysdole*
occidentalissimi illius quarentene et iuxta Edwyndich *Kynges college*   *vj acr* Egid.

19 41   Quarentena iacens in longitudine ad capud australe quarentene ultimo
dicte et est secunda quarentena de le Deddale   *incipe in parte occidentali*

:*f cχ*   ij.seliones cum.j.butta iacentes ex parte orientale ad capud eorum
australe collegii predicti quondam Johanis Redhod et sunt primi et
occidentalissimi illius quarentene et iuxta Edwyndich predictum   *j acra* Egid.
ij.seliones Prioris de bernewell iuxta et sunt curtiores aliis
selionibus antedictis ad capud eorum australe quia predicta butta iacet
in longitudine ad capud antedictum   *j acr* Egid.

mer'   xij.seliones de terra clericorum de merton iuxta et extendunt se in
longitudine per.ij.quarentenas primo selione et tribus selionibus
ultimis exceptis   *iij acr* Egid.
ij.seliones Cantarie beate marie ecclesie sancti sepulcri iuxta   *j acr* Egid

[fol. 23v.]
vij.seliones Roberti de Bluntisham quondam Henrici de Toft iuxta
mediante le Weie balk   *gylys*   *iij acr* Rotund'
j.selio Willelmi Shirwynd quondam Willelmi Tuliet iuxta   *j rod di* Rad.

sine balca	iij.seliones monialum sancte Radegunde iuxta quorum medius selio extendit se per.ij.quarentenas et est in parte forera ad capud suum australe	*j acr di* Par.
	ij.seliones hospitalis predicti iuxta quorum ultimus selio est forera et sunt ultimi et orientalissimi illius quarentene	*j acr* Egid.

20 42   Quarentena iacens in longitudine ad capud australe quarentene ultimo
dicte et est.iij.quarentena de le Dedale / / Seliones illius quarentene
debent computari ad capud eorum australe

sine balca	j.selio monialum sancte Radegunde extendens se per.ij.quarentenas et est forera ut supradictum est et primus et orientalissimus illius quarentene	*j rod* Par.
	ij.seliones Willelmi Suthwynd quondam Willelmi Tuliet iuxta	*j acr* Rad.
	j.selio Prioris de Bernewell iuxta	*di acr'* Egid.
ca b cχ	iij.seliones Collegii predicti quondam Thome de Cantebrigia iuxta	*j acra* Rad.
o	iij.seliones hospitalis predicti ocupati per clericos de merton iuxta	*j acr di* Egid.
	j.selio Cantarie beate marie ecclesie sancti sepulcri iuxta mediante le Weye balk	*di acr* Egid.
mer'	viij.seliones de terra clericorum de merton extendentes se per.ij. quarentenas ut supradictum est iuxta	*ij acr* Egid.
	j.selio Prioris de Bernewell iuxta et est ultimus et occidentalissimus illius quarentene et iuxta Edwyndych predictum	*di acr* Egid.

21 43   Quarentena ex altera parte Edwynedich predicti abuttans ad capud
suum orientale super (Ew *deleted*) Edwynedich predictum

121

THE WEST FIELDS OF CAMBRIDGE
[fol. 24r.] 24

†	*j selio quondam Nicholai moriz unacum selione curto sancti johannis*	*di acr*

mer' xiiij.curti seliones et.xx.longi seliones cum suis foreris de terra
clericorum de merton qui vocantur goidzmedole iacentes in longitudine
iuxta campum de Cotes et abuttant ad unum capud super Edwyne
Dich predictum et ad aliud capud super le clyntweye et sunt primi et
borialissimi illius quarentene   *xx acr* Egid.
ij.longi seliones cum sua forera et.vj.curti seliones eiusdem dole de
terra dictorum clericorum de merton iuxta et predicti.ij.longi seliones
abuttant ad utrumque capud ut supra et sunt longiores aliis selionibus
sequentibus in duplum   Rad.

(: †   j.selio hospitalis predicti iuxta *hic continuit(?) ad caput orientale*   *ellimin'*
super shorte furlonge usque ad berton weye versus australem   *di acr* Egid.

b   iiij.seliones monialum de bech quondam hugonis Pittok iuxta   *ij acr* Par.
         iiij.seliones Rogeri de harleston quondam Ricardi Tableter iuxta *ij acr* Egid.
inter.ij. balc' v.seliones hospitalis predicti iuxta quorum primus selio est longior
         aliis selionibus sequentibus ad capud suum orientale   *ij acr di* Egid.
         vj.seliones Rogeri de harleston quondam Ricardi Tableter iuxta *ij acr* Rotund'
         vij.seliones hospitalis predicti iuxta et sunt ultimi et australissimi
         illius quarentene et iuxta bertonweye   *j acr iij rod* Rotund'

22 🕭 44   Quarentena iacens transversa ad capud orientale.vij.selionum
         hospitali predicti ultimo dictorum et abuttat ad unum capud   *versus australem.*
         super bertonweye predictam et ad aliud capud super Edwynedich
         predictum   *versus borialem.*

cχ   ij.seliones cum sua forera Collegii predicti quondam Thome de
         Cantebrigia primi et occidentalissimi illius quarentene et primus selio
         est forera   *iij rod* Rad.

†   j.selio qui solebat esse.ij.de terra universitatis iuxta   *iij rod* Rotund.
inter.ij. balc' ij.seliones Thome de Audele iuxta *bokenham Rawlyn*   *j acr di* Par.
gun(*vile*)

mo   vj.seliones Stephani Moriz senioris quondam Willelmi Bekeswell
         iuxta   *j acr di* Rotund.
morys   v.seliones cum sua forera Johanis Pilet quondam patris sui
         iuxta   *j acr j rod* Egid.
inter.ij. balc' iij.seliones cum sua forera hospitalis predicti iuxta   *iij rod* Egid.
mo   iij.seliones Stephani Moriz senioris quondam Willelmi Bekeswell
         iuxta   *iij rod* Par.
         Sed per botolphum vendicantur   Botul' *nota*.

(: *Hi relinqui seliones istius quarentene abuttant super clynt weye versus occidentalem.*

[fol. 24v.]

nota   x.seliones Nicholai Crochman iuxta et sunt ultimi et orientalissimi
         illius quarentene et iuxta Edwynedich predictum   *ij acr* Botul'

LYTILFELD

23 45 Quarentena iacens in longitudine ad capud occidentale.vij.selionum hospitalis predicti ultimo dictorum et abuttat ad capud suum occidentale super le Clyntweye predictum *incipe prope berton' weye ex parte boriale eiusdem ex parte orientali de colys crosse.*

                v.seliones hospitalis predicti primi et australissimi illius quarentene
                et iuxta berton weye predictam   *j acr* Egid.

b   ⎧ viij.seliones cum sua forera monialum de bech quondam hugonis
Inter duas   ⎨ Pittok iuxta   *ij acr di* Par.
balcas   ⎩
.*co* c𝒳   iiij.seliones Collegii predicti quondam Thome Cantebrigia et
                vocantur mundes lond iuxta   *ij acra* Egid.
mo   iij.seliones Thome Moriz iuxta   *j acr* Egid.
                j.selio hospitalis predicti iuxta   *di acr* Egid.
                j.selio Rogeri de harleston quondam Ricardi Tableter iuxta et est
✠ *colys*   ultimus et borialissimus illius quarentene et iuxta.ij.longos seliones de
crosse   Goidzmedole ultimo dictos in quarentena de Goidzmedole   *di acr* Par.
in clynt weye

                Et sic finitur Middelfeld

[Added as a   ⎧ Summa selionum istius campi decimalium ecclesie botulphi   xxj
footnote]   ⎩ Summa selionum in istiis duobus quarentenis xxxvi seliones
                                                                                             b.s.

                              Litelfeld

                Memorandum quod garbe decimalium de omnibus selionibus iacentibus in Litilfeld qui dicuntur. Par.ut infrascriptum est debent eque partiri inter Radegundam et Egidium videlicet.j.garba Radegunde et j.garba Egidio.

                [In another hand here is placed the note intended for the end of the Middelfeld]

[fol. 25r.]

1 ♭ *46*   Quarentena inter Bertonweye predictum et bynnebrooke et abuttat ad capud suum australe versus Portebregg et est prima quarentena a Portbreg // Seliones illius quarentene debent computari ad capud eorum boriale
                Incipe ex parte australi berton weye prope ipsam et prope crucem predictam.

                  nunc v.
                vj.seliones alborum canonicorum primi et occidentalissimi   *al ca.*
                illius quarentene et iuxta bertoneweye predictam   *ij acr di* Rad.
mo   ij.seliones Stephani Moriz senioris quondam Willelmi de bokeswell
                iuxta   *j acr* Par.
mo   iiij.seliones Thome Moriz iuxta   *j acr di* Egid.
                iiij.seliones Cantarie beate marie ecclesie sepulcri   *ij acr* Egid.
                ij.seliones Johanis Pilet quondam patris sui iuxta   *morys j acr di* Egid.
                iiij.seliones Rogeri de harleston quondam Ricardi Tableter iuxta   *ij acr* Egid.

THE WEST FIELDS OF CAMBRIDGE

	mo	j.selio Thome Moriz iuxta	*di acr.* Botul' *nota*
	mo	j.selio Stephani Moriz senioris quondam patris sui iuxta	*di acr* Egid.
	mo	vj.seliones Thome Moriz iuxta	*iij acr* Egid.
		j.selio hospitalis sancti Johanis evangeliste Cantebrigie iuxta	*ij rod* Egid.
	mo	ij.seliones Thome Moriz iuxta	*iij rod* Egid.

vj.seliones Rogeri de harleston quondam Ricardi Tableter iuxta et sunt curtiores aliis selionibus antedictis ad capud eorum australe et sunt ultimi et orientalissimi illius quarentene et iuxta bynnebrook
ex parte (*occidentali* deleted) *australi et parte occidentali*   *j acr j rod* Rad.

2 ¶ 47   Quarentena iacens in longitudine ad capud boriale quarentene ultimo dicte et est secunda quarentena a pourtebregg predicto
Incipe in parte (*australi* deleted) *orientali eiusdem quarentene prope portebrigge*
ij.seliones Rogeri de harleston quondam Ricardi Tableter quorum primus selio est curtior aliis selionibus sequentibus ad capud australe et sunt primi et orientalissimi illius quarentene et iuxta bynnebrook   *j acr* Par.

mo	iij.seliones Stephani Moriz senioris quondam patris sui iuxta	*j acr* Egid.
	ij.seliones hospitalis predicti iuxta quorum decima	*j acr* hospitali predicto 1.c.

[fol. 25v.]

	mo	j.selio Stephani Moriz senioris quondam Willelmi de bekeswell iuxta	*di acr.* Par.
*nota cl*χ *G.*		j.selio Johanis barker quondam Nicholai bradenash iuxta	*di acra* Botul'
		j.selio Rogeri de harleston quondam Ricardi Tableter iuxta	*di acr* Egid.
	mo	ij.seliones Stephani Moriz senioris quondam patris sui iuxta	*iij rod* Egid.
*c*χ		iij.seliones Collegii predicti quondam Thome de Cantabrigia iuxta	*j acra j rod* Rad.
		ij.seliones hospitalis predicti iuxta	*j acr* Egid.
*c*χ *G*		j.selio Johanis barker quondam Nicholai bradenash iuxta	*iij rod* Rad.
	mo	iiij.seliones Thome Moriz iuxta	*j acr* Egid.
		ij.seliones hospitalis predicti iuxta	*j acr* Egid.
	mo	ij.seliones Stephani Moriz senioris quondam patris sui iuxta	*j acr* Egid.

ij.seliones qui solebant esse.iij.Cantarie beate marie ecclesie sepulcri iuxta   *iij rod* Egid.
iij.seliones cum.j.gora hospitalis predicti iuxta et sunt ultimi et occidentalissimi illius quarentene et iuxta Bertonweye predictam   *j acr* Egid.

3 ¶ 48   Quarentena iacens in longitudine ad capud boriale quarentene ultimo dicte et est tertia quarentena a pourtebregge predicto

j.gora et.v.seliones hospitalis predicti primi et occidentalissimi illius quarentene et iuxta Bertonweye predictam et abuttant ad capud eorum australi super cotenpath   *vj rod* Botul'

nota   iij.seliones Nicholai Crocheman iuxta   *j acr di* Egid.
vj.seliones cum.iij.buttis iacentibus ex parte occidentale ad capud eorum boriale hospitalis predicti iuxta et sunt longiores aliis selionibus antedictis nam extendunt se in longitudine a Cotenpath predicto usque ad Edwyndich   *per duas quarentenas v acr di* Egid.

LYTILFELD

sine	⎧ x.seliones Thome Audele de grauntcestre iuxta quorum.v.primi seliones extendunt se in longitudine ut supra et.v.ultimi seliones sunt curtiores aliis selionibus precedentibus ac etiam aliis selionibus sequentibus ⎨ ad capud boriale	vij acr Par.

[fol. 26r.] 26

balca †	⎧ iij.seliones Johanis Pilet quondam patris sui iuxta et extendunt se in longitudine ut supra *morys* ⎨ j.gora et.iiij.seliones de terra universitatis iuxta et extendunt se in longitudine ut supra et sunt ultimi et orientalissimi illius quarentene et iuxta bynnebrook	*ij acr* Rotund. *ij acr* Rotund.
cχ G θ	iiij.seliones Johanis Barker quondam Nicholai bradenassh iacentes in longitudine ad capud boriale dictorum selionum Thome de Audele de grauntcestre ultimo dictorum et balcati ex utraque parte et extendunt se in longitudine a capite boriali dictorum.v.selionum curtiorum usque ad Edwynedich antedictum (et sunt quarentena per se)	θ *ij acra* Rotund.
4 🖋 49	Quarentena ex altera parte Edwynedich predicti abuttans ad capud suum boriale super bertonweye et ad capud suum australe super Edwynedich et super Custesweye / / Seliones illius quarentene debent computari ad capud eorum boriale	
tylers sine balca	⎧ xj.seliones Willelmi de Thacsted primi et occidentalissimi illius ⎨ quarentene et iuxta Edwynedich ⎩ x.seliones Prioris de bernewell iuxta *nunc ix* xv.seliones eiusdem Prioris iuxta iij.seliones eiusdem Prioris iuxta	*iij acr* Rotund. *iiij acr* Rotund. *ix acr* Egid. *j acr di* Rad.
cχ G Inter.ij. balcas mo	⎧ j.selio Johanis Barker quondam Nicholai bradenassh iuxta ⎨ j.selio hospitalis predicti iuxta ⎩ j.selio Thome Moriz iuxta v.seliones Prioris de huntyngdon iuxta quorum ultimus selio est forera et sunt ultimi et orientalissimi illius quarentene	*di acr* Rad. *di acr* Egid. *di acr* Egid. *v rod.* Rad.
5 🖋 50	Quarentena abuttans ad unum capud super foreram Prioris de huntyngdon ultimo dictam et ad aliud capud super Berton weye *boriale*	

2.c.

[fol. 26v.]

nota 3	predictam / / Seliones illius quarentene debent computari ad capud eorum boriale x.buttes hospitalis predicti prime et occidentalissime illius quarentene et abuttant super foreram Prioris de huntyngdon ultimo dictam	*kx j acr* Botul'
ca sine balca ca cχ ca cχ	⎧ j.selio Johanis Pilet quondam patris sui iuxta ⎨ j.selio Collegii predicti quondam Thome de Cantabrigia iuxta ⎩ j.selio Collegii predicti quondam Willelmi Snoryng iuxta et est ultimus et orientalissimus illius quarentene et iuxta Custesweye	*di acr* Egid. *di acra* Rad. *di acra* Rad.

THE WEST FIELDS OF CAMBRIDGE

the byn broke ys the broke that the condyte goth over et so goth forth to portebreg and pasynge thatt also in quar. northwest

6 ♭ 51   Quarentena abuttans ad capud suum boriale super Custesweye predictam et ad capud suum australe super bynnebrook / / seliones illius quarentene debent computari ad capud eorum boriale   *et partim*

            ⎡ j.selio Johanis de Weston quondam Simonis de Moordon primus
sine   ⎢ et orientalissimus illius quarentene et iuxta viam que ducit a bynnebrook
balca   ⎨ ad bercariam hospitalis predicti et est curtior aliis selionibus sequentibus
            ⎢ ad capud boriale   *clar' hall di acr* Par.
c cχ   ⎢ j.selio Collegii predicti quondam Johanis de poplyngton
   di acr ⎣ iuxta   *di acra* Par. *b*
              xiiij.seliones hospitalis predicti iuxta   *ix acr* Egid.
nunc. 7.   x.seliones cum sua forera Roberti Tuliet iuxta   *cotton*   *iij acr* Rad.
              iiij.seliones Prioris de Bernewell iuxta quorum.iij.primi seliones sunt
              longiores aliis selionibus sequentibus ad capud eorum australe   *ij acr* Rad.
              v.seliones eiusdem Prioris iuxta   *ij acr di* Egid.
              viij.buttes parochialis ecclesie sancti Johanis baptiste quondam
              Ricardi Thacsted iuxta et sunt ultime et occidentalissime illius
              quarentene   *clar' hall ij acr* Rotund.

[fol. 27r.]   27

7 52   Quarentena abuttans ad capud suum occidentale super predictam viam que ducit a Bynnebrook predicto usque ad bercariam predictam et iacet in longitudine iuxta bertonweye antedictam   *incipe computare ex parte australi*

c cχ   iiij.seliones cum sua forera Collegii predicti quondam Thome de Cantabrigia quorum primus selio est forera et sunt primi et australissimi illius quarentene   *j acr et di b* Botul' nota

c cχ   ij.seliones cum sua forera eiusdem Collegii quondam Galfridi Seman iuxta   *j acr b* Par.

mo   iij.seliones cum sua forera Stephani Moriz senioris quondam patris sui iuxta   *gryfyn*   *j acr di* Egid.

       vj.seliones cum sua forera hospitalis predicti iuxta et sunt ultimi et borialissimi illius quarentene et iuxta bertonweye   *ij acr* Rad.

8 ♭ 53   Quarentena iacens transversa ad capud orientale.vj.selionum hospitalis predicti ultimo dictorum et abuttant ad capud suum boriale super bertonweye

       xiiij.seliones hospitalis predicti quorum.vj.primi et occidentalissimi seliones sunt longiores aliis selionibus sequentibus in duplum nam extendunt se in longitudine a bertonweye predicta usque ad bynnebrook predictum et primus selio est in parte forera et sunt primi et occidentalissimi illius quarentene   *vij acr* Egid.
       xij.seliones de terra Mortimer iuxta   *iij acr* Rad.

CARME FELD

iij.seliones hospitalis predicti iuxta quorum ultimus selio est longior aliis selionibus antedictis ad capud suum australe et est forera ad.vij. seliones eiusdem hospitalis iacentes ex opposito porte grangie hospitalis predicti et sunt ultimi et orientalissimi illius quarentene et iuxta viam sancti neoti *ij acr* Egid.

9 ♭ 54 Quarentena iacens transversa ad capud australe.xij.selionum de terra Mortimer ultimo dictorum et abuttant ad capud suum orientale super viam sancti Neoti predictam ex oposito grangie predicte *callyd callyd seynt Johnes ferme or bernys*

3.c.

[fol. 27v.]

vij.seliones hospitalis predicti quorum primus selio est forera et sunt primi et borialissimi illius quarentene *iij acr* Egid.
vj.seliones cum sua forera ad capud eorum orientale et cum.iij.goris iacentibus ex parte australi ad capud eorum occidentale Rogeri de harleston quondam Willelmi Ward iuxta et sunt ultimi et australissimi illius quarentene *nunc v* *ij acr* Egid.

19 ♭ 55 Quarentena iacens transversa ad capud occidentale.vj. selionum Rogeri de harleston ultimo dictorum et abuttant ad capud suum australe super Bynnebrook predictum

gun(vile) nunc.x. xj.seliones de terra Mortimer primi et orientalissimi illius quarentene. primus selio forera *iij acr* Rad.
xvj.seliones hospitalis predicti iuxta quorum.vj.primi seliones sunt longiores aliis selionibus sequentibus in duplum nam extendunt se in longitudine a bynnebrook predicto usque ad bertonweye predictam *viij acr* Egid.

b.c. cχ iij.seliones cum sua forera ad capud eorum australe Collegii predicti quondam Thome de Cantabrigia iuxta et sunt ultimi et occidentalissimi illius quarentene et iuxta viam que ducit a bynnebrook predicto usque ad bercariam antedictam *ij acr b* Botul.

Et sic Finitur Lytilfeld

summa selionum istius campi decimalium ecclesie botulphi xxvj [Added as a footnote]

Memorandum quod garbe decimalium de omnibus selionibus iacentibus in le Carmefeld qui dicuntur. Par.ut infrascriptum est debent eque partiri inter Radegundam et Egidium videlicet.j.garba Radegunde et.j.garba egidio exceptis garbis decimalium de.ij. selionibus Ricardi Martyn et de.v. selionibus Ricardi de Ardern similiter iacentibus

Carme Feld

[fol. 28r.]

in Eldenewenham croft que debent eque partiri inter Radegundam et domum sancti Petri / et exceptis garbis decimalium de.v.selionibus Collegii corporis christi quondam Thome de Cantabrigia iacentibus

THE WEST FIELDS OF CAMBRIDGE

iuxta le longebalk que debent eque partiri inter Egidium et Botulphum ut patet infra *Old Newenham weye commyth from the crosse that standes in the wey from cant' to granceter et ledyth to portebrygge*

1 ♭ 56 Quarentena abuttans ad capud suum boriale super Eldenewenham et ad capud suum australe super Mortimeres madewe *incipe per inclausum mortimer' prope ripam ex parte (australi* deleted) *predicti inclausi (————)occidentali nunc in tenura cant' ville et gylis occupat*
v.buttes hospitalis sancti Johanis evangeliste Cantabrigie primi et
orientalissimi illius quarentene   *ij acr di* domui sancti Petri
ij.seliones Ricardi Martyn quondam Johanis Martyn
iuxta   *readde*   *j acr* Par. inter Rad. et Petri
v.seliones Ricardi de Ardern quondam Matilde hechyn iuxta
rede chapman   *ij acr di* Par. inter Rad. et Petri

nota   ij.seliones dicti Ricardi quondam dicte Matilde iuxta *rede chapman*   *j acr* Botul.
j.selio hospitalis predicti iuxta   *di acr* domui sancti Petri
nota mo   ij.seliones Thome Moriz iuxta   *j acr* Botul.
ij.seliones hospitalis predicti qui vocantur le lampeaker iuxta
                      j acr domui sancti Petri

sine balca ⎧ j.selio Ricardi Martyn quondam Johanis Martyn
     ⎨ iuxta *rede*   *di acr* domui sancti Petri
nota   ⎩ j.selio Ricardi Foukes quondam patris sui iuxta et est ultimus et
occidentalissimus illius quarentene   *di acr* Botul.

.2. ♭ 57 Quarentena abuttans ad capud suum orientale super Eldenewenham *weye mediante alta via vocata olde newenham iacet ex parte occidentali campi predicti.*
Seliones illius quarentene debent computari ad capud eorum occidentale

✠   ⎧ ij.seliones henrici de bech primi et australissimi illius quarentene et
sine balca ⎨ iuxta Eldenewenhamwey   *j acr* domui sancti Petri
rede   ⎩ ij.seliones qui solebant esse.iij.Ricardi martyn quondam Andree
hardy iuxta   *j acr* d.s.p.
G   *cχ*   j.selio Johanis touche quondam Johanis Martyn iuxta   *di acr* d.s.p. G
ij.seliones hospitalis predicti iuxta   *j acr* d.s.p.

                          4.c.

[fol. 28v.]

vj.seliones Galfridi Wardeboys iuxta et sunt curtiores aliis
selionibus antedictis ad capud eorum orientale   *ij acr iij rod* d.s.p.
nota   j.selio Nicholai Crocheman iuxta *rede*   *iij rod* Botul.
rede   ij.seliones qui solebant esse.iij.Ricardi Martyn quondam Andree
dufowse hyll   hardy iuxta   *j acr* d.s.p.
    ⎧ j.selio hospitalis predicti iuxta et est longior aliis selionibus antedictis
    ⎪ ad capud orientale   *j rod* d.s.p.
s cχ ⎨ j.selio ocupatus per Ricardum Martyn iuxta   *j rod c* d.s.p.
    ⎪
inter.ij. balcas ⎨ j.selio Johanis Sharp quondam Thome de Wyntringham iuxta   d.s.p. ⎫
   † ⎩ j.selio Roberti Martyn quondam patris sui iuxta   d.s.p. ⎬ *di acr*
j.selio hospitalis predicti iuxta   *di acr* d.s.p.

CARME FELD

[*lxiij*] *c*χ inter.ij. balc' *in manibus* *clar' hall*	⎧ ij.seliones Collegii predicti quondam Thome de Cantebrigia iuxta ⎨ ij.seliones Roberti Martyn quondam patris sui iuxta ⎩ ij.seliones Johanis Pilet quondam patris sui iuxta		*j acra* d.s.p. (*hospitalis*) *di acr* d.s.p. (*morys*) *j acr* d.s.p.
sine balca nota	⎧ j.selio Stephani Moriz senioris quondam patris sui iuxta ⎩ j.selio Nicholai Crocheman iuxta et est ultimus et borialissimus illius quarentene et iuxta Frosshlake weye *ys the lane commynge up from Newenham towards cotenpath levynge yt on the lyfte hand as you com up from Cambreg' turnyng yt ryght weste*		*di acr* d.s.p. *di acr* Botul'
3 ♭ 58	Quarentena extendens in longitudine ultra viam que ducit a Frosshlake usque ad pourtebregg et abuttat ad capud suum boriale super cotenpath / / Seliones illius quarentene debent computari ad capud eorum boriale		
nota sine balca *clar' hall* *clar' hall* *mo* *b*	⎧ j.selio hospitalis predicti primus et orientalissimus illius quarentene ⎨ et iuxta le longebalk ⎩ j.selio Johanis de Weston quondam Roberti de Comberton iuxta j.selio qui solebat esse.ij.Rogeri harleston quondam Willelmi Ward iuxta v.seliones Johanis de Weston quondam Roberti de Comberton iuxta j.selio Stephani Moriz senioris qui vocatur Shermannisrod iuxta j.selio monialum de bech iuxta		*j rod di* Botul' *j rod di* Egid. *di acr* Egid. *ij acr* Egid *j rod* Rad. *j rod* Par.

29

[fol. 29r.]

mo (*hospitalis*)	j.selio Stephani Moriz senioris quondam patris sui iuxta vj.seliones Andree de Cotenham quondam Johanis de Toft iuxta *et predicti xiij iacent inter duas balcas* ij.seliones Cantarie beate marie ecclesie sancti petri extra trumpintongates iuxta j.selio henrici de bech iuxta j.selio cantarie beate marie ecclesie sancti petri extra trumpintongates iuxta		*j rod* Egid. *ij acr di* Rad. *j acr* Egid. *di acr* Par. *di acr* Egid.
mo	ij.seliones Stephani Moriz senioris quondam patris sui iuxta j.selio Prioris de huntyngdon iuxta j.selio cantarie beate marie ecclesie sancti petri extra trumpintongates iuxta		*j acr* Egid. *di acr* Rad. *di acr* Egid.
inter.ij. balcas	⎧ ij.seliones Johanis Pilet quondam patris sui iuxta *morys* ⎨ j.selio Andree de Cotenham quondam Johanis de Toft iuxta *hospit'* ⎩ j.selio Rogeri de harleston quondam Willemi Ward iuxta et iuxta custesbalk		*j acr* Egid. *di acr* R ad *di acr* Botul. nota
custes balk'——custes balke			
*g c*χ *g c*χ	⎧ iij.seliones Collegii predicti quondam Henrici Tanglemere et quondam ⎨ Roberti de Comberton iuxta et iuxta custesbalke predictam Par. ⎨ ij.seliones eiusdem Collegii quondam dicti henrici et quondam ⎩ dicti Roberti iuxta Egid.	⎫ ⎬ *ij acra* :: ⎭	
	vj.seliones qui solebant esse.ix.Rogeri harleston quondam Willelmi Ward iuxta		*iij acr* Par.

THE WEST FIELDS OF CAMBRIDGE

	j.selio hospitalis predicti iuxta	*di acr* Egid.
inter.ij. balc'	⎧ j.selio Prioris de huntyngdon iuxta	*di acr* Rad.
clar' hall	⎪ iij.seliones Johanis de Weston quondam Roberti de Comberton iuxta	Egid ⎫
	⎨ iiij.seliones eiusdem Johanis quondam Johanis de Comberton iuxta	Rad. ⎬ *iij acr di*
mo	⎧ iiij.seliones Stephani Moriz senioris quondam patris sui iuxta	*ij acr* Egid.
Inter.ij. balcas	⎨ j.selio hospitalis predicti iuxta	*j rod* Egid.
mo	vj.seliones.j.gore et.ij.buttes Ricardi Moriz iuxta et sunt ultimi et occidentalissimi illius quarentene et iuxta bynnebrook	*iij acr di* Egid.

4 ▸ 59 Quarentena iacens transversa ad capud australe.vj.selionum Ricardi Moriz ultimo dictorum et abuttat ad capud suum occidentale versus Pourtebregg / / Seliones illius quarentene debent computari ad capud eorum orientale

5.c.

[fol. 29v.]

mo	⎧ iij.seliones cum sua forera Stephani Moriz senioris quondam patris sui primi et borialissimi illius quarentene et primus selio est forera	*ij acr* (deleted) *j acr di* Egid.
sine balca	⎨ j.selio dicti Stephani quondam Johanis de berton iuxta	*di acr* Botul.
nota mo	⎪ j.selio hospitalis predicti iuxta	*di acr* Rad.
	⎩ iij.seliones eiusdem hospitalis iuxta	*j acr di* Par.
	⎧ j.selio Rogeri de harleston quondam Ricardi Tableter iuxta	*di acr* Egid.
o cχ	⎪ j.selio Johanis barker quondam Nicholai bradenassh iuxta	Rad. ⎫ *j acr et*
sine balca	⎨ ij.seliones eiusdem Johanis quondam dicti Nicholai iuxta	Egid. ⎬ *di G*
o	⎩ j.selio hospitalis predicti iuxta et est ultimus et australissimus illius quarentene et iuxta Eldenewenham Weye	*di acr* Egid.

.5. ▸ 60 Quarentena iacens transversa ad capud orientale quarentene ultimo dicte et abuttat ad capud suum australe super eldenewenhamweye predictam

	j.selio hospitalis predicti primus et occidentalissimus illius quarentene et est forera	*di acr* Egid
	j.selio Ricardi Martyn quondam Roberti de Comberton iuxta *rede*	*di acr* Egid.
(fan)		
g	⎧ *cχ nunc* xiiij.seliones cum sua forera Collegii predicti quondam Thome de	
	.13. Cantebrigia iuxta quorum primi seliones sunt longiores aliis selionibus	
k	⎨ sequentibus ad capud boriale	*iij acr j rod* Rad.
g	⎩ *cχ nunc* iiij.seliones cum sua forera eiusdem Collegii quondam Johanis de	
	.3. Poplyngton iuxta et iuxta Custisbalk *in toto vij acr*	*ij acr* Par.
lovel	⎧ ij.seliones Thome Moriz iuxta mediante custisbalke predicte	*j rod di* Egid.
inter.ij. balcas	⎨ vij.seliones eiusdem Thome iuxta *lovel*	*j acr di* Par.
rede	⎩ ij.seliones cum sua forera Ricardi Martyn quondam Roberti Bolour iuxta	*j acr* Rad.
nota	iiij.seliones cum sua forera hospitalis predicti iuxta	*ij acr* Botul.
clar' hall	⎧ j.selio Johanis de Weston quondam Simonis de Moorden iuxta	*j rod* Par.
inter.ij. balcas	⎨	
$ 8 *cχ*	⎩ j.selio Collegii predicti quondam Johanis de poplyngton	*j rod* Par.

130

CARME FELD

clar' hall	j.selio Cantarie beate marie ecclesie sancti sepulcri iuxta	*iij rod* Egid.
	iij.seliones cum sua forera Johanis de Weston quondam Roberti Comberton iuxta	*j acr di* Egid.

[fol. 30r.]   30

mo	j.selio cum sua forera Thome Moriz iuxta et iuxta le longebalke	*di acr* d.s.p.
g cχ	⎧ v.seliones qui solebant esse.vij.cum sua forera Collegii predicti	
Inter.ij. balcas	⎪ quondam Thome de Cantebrigia iuxta mediante le longebalk predicta	
nota	⎨ et sunt longiores aliis selionibus antedictis ad capud eorum boriale / /	
rede	⎪ partiti inter Egidium et Botulphum *ij acr*	Par. inter Egid. et Bot.
	⎩ iiij.seliones cum sua forera Ade de Kynxton quondam patris sui iuxta	*ij acr* Egid.
nota	iiij.seliones cum sua forera Roberti Barbour quondam Roberti Bolour	
o cχ	modo Collegii iuxta	*fam ij acr* Botul.
nunc 5	vj.seliones cum sua forera Johanis Barker iuxta quorum .iij.primi seliones	
o cχ	quondam Nicholai Bradenassh et.iij.ultimi seliones quondam Johanis Branton iuxta	G. *iij acr* d.s.p.
	ij.seliones cum sua forera modo Collegii Roberti Barbour quondam Roberti Bolour iuxta	*fam i acra* d.s.p.
mo	⎧ iij.seliones cum sua forera Stephani Moriz senioris quondam patris sui iuxta	*j acr di* Par.
sine balca	⎨ iij.seliones cum sua forera Ricardi de Arderne quondam Ricardi Tableter iuxta et sunt ultimi et orientalissimi illius quarentene et	
	⎩ ultimus selio est forera *blackbol*	*j acr di* Par.

6 ♭ 61	Quarentena iacens transversa ad capud boriale.iij.selionum Ricardi de Ardern' ultimo dictorum / / Seliones illius quarentene debent computari ad capud eorum orientale

sine balca	⎧ j.selio Nicholai Crocheman.primus et australissimus illius quarentene et est forera	*di acr* Egid.
	⎩ ij.seliones Johanis Pilet quondam patris sui iuxta *morys*	*j acr* Egid.
†	ij.seliones de terra universitatis iuxta	*j acr j acr* Rotund.
s cχ o	j.selio Collegii predicti quondam Johanis de poplyngton iuxta	*j acr* Rotund.
clar' hall	⎧ j.selio Johanis de Weston quondam Simonis de Mordon iuxta	*iij rod* Rotund.
	⎪ Seliones predicti sunt longiores aliis selionibus sequentibus magis	
sine balca	⎨ quam in duplum nam extendunt se in longitudine a newenham croft usque ad le longebalke	
	⎩ vj.seliones Nicholai Crocheman iuxta et iuxta Frosshelakeweye	*ij acr di* Egid.

                                                                              6.c.

[fol. 30v.]

a	j.gore et.v.seliones cum sua forera ad capud eorum orientale dicti Nicholai Crocheman iuxta mediante Frosshelake Weye predicta et predicta gore et.v.seliones sunt quasi.j.quarentena per se sicut ibidem iacent *morys*	*j acr* Egid.
a	j.curta forera hospitalis predicti iacens transversa ad capud orientale.vj.selionum Nicholai Crocheman ultimo dictorum et est	
∴	quasi forera ad eosdem *mediante longebalke*	*j rod* Egid.

131

THE WEST FIELDS OF CAMBRIDGE

	a	j.alia forera eiusdem hospitalis extendens se in longitudine ultra Frosshelakeweye predictum et iacet transversa ad capud occidentale.vj. et.v.selionum Nicholai Crocheman ultimo dictorum et est quasi forera ad eosdem *mediante longebalke*	Egid.
7 ♄ 62		Quarentena abuttans ad capud australe super Cotenpath predictam // Seliones illius quarentene debent computari ad capud eorum australe	
sine balca	⎧ ⎨ ⎩	j.selio Willelmi Aleyn de bokesworth primus et orientalissimus illius quarentene et iuxta le longebalke *hospitalis* j. selio hospitalis predicti iuxta	*di acr* Par. *j rod* Egid.
.*b*.		j.selio monialum de bech iuxta	*di acr* Par.
inter.ij. balcas	⎰ ⎱	iiij.seliones Ricardi Martyn quondam Rogeri Russell iuxta *rede* j.selio henrici de beche iuxta	*ij acr* Rad. *di acr* Par.
.*b*.		j.selio monialum de beche iuxta	*di acr* Par.
inter.ij. balcas *mo*	⎰ ⎱	ij.seliones Stephani Moriz senioris quondam patris sui iuxta quorum secundus selio est in parte forera ad capud suum boriale	*j acr di* Egid.
		vij.seliones et.ij.gores cantarie beate marie ecclesie sancti Petri extra trumpyntongates qui vocantur le goredaker iuxta et sunt curtiores aliis selionibus antedictis ad capud eorum boriale	*ij acr di* Par.
$ *cχ (cl)*		j.selio Collegii predicti quondam Thome de Cantebrigia iuxta	*di acr* Rad.
Inter	⎰	j.selio Ricardi Martyn quondam Rogeri Russell iuxta *rede*	*j rod* Egid.

[fol. 31r.]   31

ij. balcas *mo* *mo*	⎰ ⎱	iij.seliones Stephani Moriz senioris quondam patris sui iuxta j.selio Thome Moriz iuxta	*j acr di* Egid. *di acr* Egid.
.*b*. inter.ij. balcas	⎰ ⎱	j.selio monialum de bech iuxta j.selio henrici de bech iuxta	*di acr* Par. *j rod* Par.
∴ *cχ* .xx.		iij.seliones Collegii predicti quondam Thome de Cantebrigia iuxta et iuxta custesbalk	*j acra b* Rad.
inter.ij. balcas clar' hall	⎧ ⎨ 	j.selio hospitalis predicti iuxta mediante custisbalke predicta et est longior aliis selionibus antedictis ad capud suum boriale j.selio Johanis de Weston quondam Roberti de Comberton iuxta	*iij rod* Rad. *di acr* Egid.
.*b*.		j.selio monialum de bech iuxta	*di acr* Par.
$ *cχ* Inter.ij. balcas .*b*.	⎨ 	j.selio Collegii predicti quondam poplyngton *j rod et di (cl)* Par. *di acr* j.selio Johanis de Weston quondam Simonis de Mordon iuxta iuxta *(clar' hall) di acr* Par. j.selio monialum de bech iuxta est in parte forera ad capud suum boriale	 *di acr* Par.
		j.selio hospitalis predicti iuxta et est curtior aliis selionibus antedictis ad capud boriale	*di acr* Egid.
		vj.seliones Stephani Moriz senioris quondam patris sui iuxta	*nunc 4ᵒʳ ij acr* Egid.
†		iv.seliones de terra universitatis iuxta	*j acr* Rotund.
		j.selio hospitalis predicti iuxta	*di acr* Egid.
		j.selio Ricardi Martyn quondam Roberti de Comberton iuxta *rede*	*di acr* Egid.
		ij.seliones hospitalis predicti iuxta	*di acr* Egid.
$ *cχ cl*		j.selio Collegii predicti quondam Thome de Cantebrigia iuxta *di acra*	*iij rod* Rad.
		j.selio hospitalis predicti iuxta	*di acr* Rad.

6 cχ cl		ix.seliones tamen aliqui eorum sunt buttes cum sua forera qui vocantur godynys buttes Collegii predicti quondam Johanis godyn capellani iuxta et sunt ultimi et occidentalissimi illius quarentene et iuxta bynnebrook	ij acr di Rad.
8 ♭ 63		Quarentena continens.xij.seliones iacens transversa ad capud boriale. iij. selionum Collegii predicti quondam Thome de Cantebrigia ultimo dictorum ad tale signum.xx.et abuttant ad capud suum occidentale super Custisbalke predictum	†

[fol. 31v.]

sine balca	⎧ ⎨ ⎩	vj.seliones hospitalis predicti primi et australissimi illius quarentene et primus selio est forera	ij acr Egid.
		vj.seliones Rogeri de harleston quondam Ricardi Tableter iuxta et sunt ultimi et borialissimi illius quarentene et ultimus selio est forera	j acr di Par.
9 ♭ 64		Quarentena iacens transversa ad capud boriale.ix.selionum Collegii predicti ultimo dictorum qui vocantur godines buttes et abuttat ad capud suum occidentale super bynnebrook predictum	
†		iij.seliones de terra universitatis primi et australissimi illius quarentene et primus selio est forera et abuttant ad capud eorum occidentale super bynnebrook et ad capud eorum orientale super foreram monialum de bech	j acr di Rad.
Inter.ij. balcas	⎧ ⎨ ⎩	iiij.seliones Rogeri de harleston quondam Willelmi Ward iuxta quorum primus selio abuttat ad utrumque capud ut supra et.iij. ultimi seliones sunt longiores aliis selionibus antedictis ad capud eorum orientale et abuttant super custesbalke ad capud predictum et ad capud eorum occidentale super bynnebrooke et primus selio dictorum.iij.selionum est in parte forera ad capud suum orientale	ij acr Egid.
mo		ij.seliones Stephani Moriz senioris quondam patris sui iuxta	j acr Egid.
clar' hall		j.selio Johanis de Weston quondam Johanis de Comberton iuxta	di acr Rad.
$ cχ	⎧	j.selio Collegii predicti quondam Thome de Cantebrigia iuxta (cl)	iij rod Rad.
Inter.ij. balcas	⎨		
.b.	⎩	j.selio monialum de bech iuxta	di acr Par.
mer		vj.seliones de terra clericorum de merton iuxta nunc 5	ij acr Egid.
.b.		ij.seliones monialum de bech iuxta	j acr Par.
mo		j.selio Stephani Moriz senioris quondam patris sui iuxta	di acr Egid.
clar' hall	⎧	j.selio Johanis de Weston quondam Johanis de Comberton iuxta	di acr Rad.
Inter.ij. balcas	⎨		
mo	⎩	j.selio Ricardi Moriz iuxta	di acr Egid.
$ cχ		j.selio Collegii predicti quondam Thome de Cantebrigia iuxta di acra (cl)	Rad.
mo		iij.seliones Ricardi Moriz iuxta	(di deleted) j acr Egid.
o cχ		iij.seliones Collegii predicti quondam Galfridi Seman iuxta	
			cum sequentibus vj rod Egid.

[fol. 32r.]

viij.seliones eiusdem Collegii quondam dicti Galfridi iuxta et sunt
ultimi et borialissimi illius quarentene et iuxta bynnebrook *isti xj
seliones iacent transversi terras eiusdem quarentene predicte*   vj rod (cl) Par.
                                            cum duobus precedentibus iij acr

THE WEST FIELDS OF CAMBRIDGE

10 ℞ 65	Quarentena ex altera parte de Custesbalke predicta abuttans ad capud suum boriale super bynnebrook predictum / / Seliones illius quarentene debent computari ad capud eorum boriale		
	iij.seliones Johanis de Weston primi et occidentalissimi illius quarentene et iuxta custesbalk predictum et abuttant super bynnebrook	*clar' hall*	*j acr* Egid.
	⎡j.selio dicti Johanis quondam Ele de bekeswell iuxta et abuttat ut supra		
		clar' hall	*di acr* Par.
sine balca	⎨iij.seliones dicti Johanis quondam Rogeri de Costeseye iuxta et abuttant ut supra	*clar' hall*	*j acr* Egid.
	⎣j.selio hospitalis predicti iuxta et abuttat ut supra		*di acr* Egid.
tylers	⎡j.selio Willelmi Aleyn de bokesworth iuxta et abuttat ut supra		*di acr* Par.
Inter.ij.balc'	⎨j.selio Johanis de Weston quondam Johanis Martyn iuxta et abuttat ut supra	*clar' hall*	*di acr* Par.
†	ij.seliones de terra universitatis iuxta et abuttant ut supra		*j acr* Rotund.
mo	j.selio Stephani Moriz senioris quondam Roberti de brygham et est longior aliis selionibus sequentibus in duplum et abuttat ut supra		*di acr* Egid.
nota	ij.seliones hospitalis predicti iuxta et abuttat ut supra		*j acr* Botul.
c cχ	ij.seliones Collegii predicti quondam Thome de Cantebrigia iuxta et abuttant ut supra		*j acra* ᵇ Rad.
.ᵇ.	iij.seliones monialum de bech iuxta et abuttat ut supra		*vj rod* Par.
c cχ	ij.seliones Collegii predicti quondam Galfridi Seman iuxta et abuttant ut supra	*nunc unus*	*j acra* ᵇ Par.
clar' hall	iij.seliones Johanis de Weston quondam Rogeri de Costiseye iuxta et iuxta le longebalke et abuttant ut supra		*vj rod* Egid.
mo	⎡ij.seliones Thome Moriz iuxta mediante le longebalke predicta et		
Inter.ij. balcas	⎨ sunt curtiores aliis selionibus antedictis ad capud eorum australe et abuttant ut supra		*j acr* Egid.
	⎣iij.seliones qui solebant esse.vj.Prioris de huntyngdon iuxta et abuttant ut supra		*ij acr* Rad.
c cχ	⎡j.selio Collegii predicti quondam Thome de Cantebrigia iuxta et abuttat ut supra		*di acra* ᵇ Rad.
Inter.ij. balcas	⎨j.selio Thome Moriz iuxta et abuttat ut supra		*di acr* Egid.
mo	⎣j.selio Johanis de Weston quondam Johanis de Comberton	*clar' hall*	*j acr*

[fol. 32v.]

	iuxta et est valde latus ad capud suum boriale et abuttat ut supra		Rad.
	ij.seliones Roberti de brygham qui vocantur gagges aker iuxta et abuttant ut supra		*j acr* Egid.
j acra *cχ*	ij.seliones Collegii predicti quondam Galfridi Seman iuxta et abuttant ut supra	ᵇ	Egid.
	ij.seliones Prioris de huntyngdon iuxta quorum primus selio abuttat ut supra et secundus selio abuttat in parte super foreram hospitalis predicti		*j acr* Rad.
gunvile rawlyns	⎡j.selio Thome de Audele iuxta et abuttat super foreram predictam *bukenham*		*di acr* Par.
$ *cχ* Inter.ij. balcas	⎨j.selio Collegii predicti quondam Thome de Cantebrigia iuxta et abuttat super foreram predictam	*(cl)*	*j rod* Rad.
	⎣j.selio hospitalis predicti iuxta et abuttat super foreram predictam		*j rod* Egid.
$ *cχ* *vij rodes*	iij.seliones Collegii predicti quondam Galfridi Seman iuxta et abuttant super foreram predictam	*(cl)*	Par.
	iiij.seliones Prioris de huntyngdon iuxta et abuttant super foreram predictam		*ij acr* Rad.

CARME FELD

	mo	ij.seliones Thome Moriz iuxta et abuttant super foreram predictam *(lovel)*	*j acr* Rad.
$	*cχ* *di acra*	j.selio Collegii predicti quondam Willelmi Snoryng iuxta et abuttat super foreram predictam	*(cl)* Rad.
		vj.seliones Rogeri de harleston quondam Ricardi Tableter iuxta abuttantes super dictam foreram	*ij acr* Egid.
c iij rod cχ *j acra*		j.selio Collegii predicti quondam Galfridi Seman iuxta et abuttat super foreram predictam	*b* Par.

clar' hall ⎡ j.selio Johanis de Weston quondam Simonis de Mordon iuxta
Inter.ij. balcas ⎢ abuttans super dictam foreram   *di acr* Par.
$ *di acra cχ* ⎨ j.selio Collegii predicti quondam Johanis poplyng iuxta et abuttat
              ⎣ super foreram predictam   *(cl)* Par.

†   iij.seliones de terra universitatis iuxta et abuttant super foreram predictam   *iij rod* Rotund'

$ *di acra cχ*   j.selio Collegii predicti quondam henrici Tanglemere iuxta abuttans
l iij rod   super dictam foreram   *(cl)* Egid.

    ⎡ v.seliones Henrici de bech iuxta et abuttant super foreram predictam
    ⎢ 2° *K.Coll*   *ij acr* Par.
sine balca ⎨ j.selio Thome Sturmyn quondam Willelmi atte cundut iuxta et est
regal' coll' ⎢ curtior aliis selionibus antedictis ad capud suum boriale et ultimus et
    ⎣ orientalissimus illius quarentene et iuxta le longe grene *di acr* Rad.

11 ☙ 66   Quarentena iacens transversa ad capud boriale.v.selionum Henrici de
    bech ultimo dictorum et abuttat ad unum capud super le longe grene
    predictum et ad aliud capud super bynnebrook predictum

    v.seliones cum sua forera hospitalis predicti primi et australissimi

[fol. 33r.]   33

                illius quarentene quorum primus selio est forera   *ij acr di* Rad.
c cχ   vij.seliones qui solebant esse.xj.Collegii predicti quondam Galfridi
      Seman iuxta quadam lata balca mediante   *iij acr di b* Rad.
      ix.seliones Prioris de hungtyngdon iuxta   *iij acr'* Rad.
cχ   ix.seliones cum sua forera qui solebant esse.xij.Collegii predicti quondam
      Galfridi Seman iuxta et sunt ultimi et borialissimi illius quarentene et
      iuxta bynnebrook   *iiij acr b*. Rad.*.:*

.12. ☙ 6   Quarentena iacens in longitudine iuxta le longe grene predictum ex
      oposito ecclesie sancti Johanis in melnestrete // Seliones illius quarentene
      debent computari ad capud **eorum boriale** ab[*uttat*] *in capite australi super*
      mortymeres close
      *Incipe in parte (*occidentali* deleted) *australi porte coll' regal'*.

        mer'   xvj.seliones cum sua curta forera ad capud eorum australe de terra
      clericorum de merton primi et orientalissimi illius quarentene et
      iuxta le longe grene predictum   *v acr'* Egid.
c $ *cχ* ⎧ j.selio Collegii predicti quondam Henrici Tanglemere iuxta   *di acra (cl)* Rad.
Inter.ij. balcas ⎨ iij.seliones Johanis Pilet quondam patris sui iuxta *Morys*   *j acr* Egid.
   mo   iiij.seliones Thome Moriz iuxta   *ij acr* Egid.

THE WEST FIELDS OF CAMBRIDGE

.b.	j.selio monialum de bech iuxta et est longior aliis selionibus antedictis et in parte forera ad capud suum australe	*iij rod* Par.
c cχ	iij.seliones Collegii predicti quondam Thome de Cantebrigia iuxta	*j acra et di* b Rad.
mo	iij.seliones Thome Moriz iuxta *lovel*	*j acr di* Egid.
c cχ	ij.seliones Collegii predicti quondam Galfridi Seman iuxta ⎫	*iij acra* b Egid.
c cχ	iiij.seliones Collegii predicti quondam eiusdem Galfridi iuxta ⎭	b Par.
Inter.ij. balcas	⎧ ij.seliones qui solebant esse.iiij.Johanis Cotton quondam Thome Comberton iuxta	*j acr* Rotund'
	⎩ ij.seliones Rogeri de harleston quondam Ricardi Tableter iuxta	*j acr* Egid.
mo	j.selio Stephani Moriz senioris quondam patris sui iuxta	*di acr* Egid.
Inter.ij. balcas	⎧ j.selio Rogeri de harleston quondam Ricardi Tableter iuxta	*di acr* Egid.
	⎨ ij.seliones Johanis de Weston quondam Rogeri de Costyseye iuxta	
	⎩ et iuxta le longebalke predictam *clar' hall*	*j acr* Egid.

[fol. 33v.]

mo	iij.seliones Stephani Moriz senioris quondam patris sui iuxta mediante le longe balke predicta et sunt curtiores aliis selionibus antedictis ad capud eorum boriale	*ij acr* Egid.
Inter.ij. balcas	⎧ j.selio Johanis de Weston quondam Johanis Martyn iuxta *clar hall*	*di acr* Rad.
.b.	⎩ j.selio monialum de bech iuxta	*di acr* Par.
mo	iij.seliones Stephani Moriz senioris quondam patris sui iuxta	*j acr* Egid.
	j.selio hospitalis predicti iuxta et est in parte forera ad capud suum australe	*di acr* Egid.
di acra cχ c	j.selio Collegii predicti quondam Thome de Cantebrigia iuxta et est curtior aliis selionibus antedictis ad capud suum australe et ultimus et occidentalissimus illius quarentene et iuxta unum selionem Stephani Moriz senioris quondam Roberti de brygham extendentem se in longitudine usque ad bynnebrook predictum	b Rad.
13. (4) 68	Quarentena iacens transversa ad capud australe.ij.selionum Johanis de Weston ultimo dictorum ad tale signum et abuttat ad capud suum occidentale super le longebalke predictam et ad capud suum orientale super le Carmedole furlong	
clar' hall sine balca .b.	⎧ iij.seliones Johanis de Weston quondam Johanis Amy de hichin primi et borialissimi illius quarentene et primus selio est forera	*j acr* Rad.
	⎩ ij.seliones monialum de bech iuxta	*iij rod* Par.
$ cχ(cl)	j.selio Collegii predicti quondam Galfridi Seman iuxta	*di acra iij rod* Par.
clar' hall	iij.seliones Johanis de Weston quondam Rogeri de Costiseye iuxta quorum ultimus selio est longior aliis selionibus antedictis ad capud suum orientale	*j acr* Egid.
Inter.ij. balc' *clar' hall*	⎧ j.selio ocupatus per Andream de Cottenham iuxta *hospitalis*	*j rod* Rad.
	⎨ j.selio ocupatus per Priorem de huntyngdon iuxta	*j rod* Rad.
	⎩ j.selio (monialum *deleted*) ocupatus per Johanem de Weston iuxta	*j rod* Rad.
†	iij.seliones de terra universitatis iuxta	*j acr di* Rotund'
mo	⎧ j.selio Thome Moriz iuxta *lovel*	*j rod di* Egid.
inter.ij. balc' *buckenham*	⎨ j.selio Ricardi Martyn junioris quondam Johanis de Comberton iuxta	*j rod* Rad.

CARME FELD

[fol. 34r.]

nota		j.selio hospitalis predicti qui vocatur barkeresaker iuxta	*iij rod* Botulph.
Inter.ij. balcas		j.selio Johanis Pilet quondam patris sui iuxta morys	*j acr* Egid.
mo		iij.seliones Stephani Moriz senioris quondam Rogeri de brandon	
sine balca		iuxta et sunt longiores aliis selionibus sequentibus fere in duplum et ultimus selio est forera in medio	*ij acr di* Rad.
mo		ij.seliones Thome Moriz iuxta	*di acr* Egid.
		j.selio hospitalis predicti iuxta	*j rod* Egid.
ꝋ		iiij.seliones cum sua forera ad capud eorum occidentale Johanis Pilet quondam patris sui iuxta et iuxta Frosshlakeweye predictam *morys*	*j acr di* Egid.
		+4ᵒʳ *nunc*	
a'		vj.seliones eiusdem Johanis Pilet quondam patris sui iuxta mediante	
Idem		Frosshelakeweye predicta et extendunt se in longitudine a le longebalke predicta usque ad foreram hospitalis predicti extendentes se in longitudine ultra Frosshelakeweye predictum ut supradictum est et sunt quasi.j.quarentena per se sicut iacent	*j acr' di* Egid.
a'		iij.buttes cum sua forera predicti Johanis Pilet quondam patris sui iacentes transverse ad capud orientale.iiij.selionum eiusdem Johanis Pilet ultimo dictorum et abuttant ad capud eorum australe super Frosshelakeweye predictam	Egid.
		Nunc verse sunt curt' 4ᵒʳ selionibus eiusdem Johis Pylett ad talem signum ꝋ	
14 ℔ 69		Quarentena vocata Carmesdole Furlong // Seliones illius quarentene debent computari capud eorum occidentale	
sine balca		xx.seliones cum suis foreris qui vocantur Carmedole hospitalis predicti primi et australissimi illius quarentene et iuxta Frosshelakeweye predictam et abuttant ad capud eorum orientale super tenementum et pratum domini Roberti Mortimer et ultimus selio est in parte forera ad capud predictum	*xj acr* Egid.
$ cχ		ij.seliones cum.j.lata balca iacentes ex parte australe ad capud eorum occidentale Collegii predicti quondam Galfridi Seman iuxta et sunt longiores aliis selionibus antedictis *Whelpes acre* (cl)	*j acra et j rod* Egid

[fol.34 v.]

		ad capud predictum et curtiores aliis selionibus antedictis ad capud eorum orientale et abuttant super foreram hospitalis predicti ad capud eorum orientale	Par.
Inter.ij. balc'		j.selio Rogeri de harleston quondam Ricardi Tableter iuxta et est latus ad capud suum orientale et abuttat super foreram predictam	*di acr'* Egid.
		j.selio Thome Moriz iuxta et abuttat super foreram predictam	*di acr'* Egid.
$ *j rod* cχ		j.selio Collegii predicti quondam Galfridi Seman iuxta et abuttat super foreram predictam	(cl) Par.
mo		j.selio Thome Moriz iuxta et abuttat super foreram predictam	*di acr* Egid.
✻		iij.seliones hospitalis predicti iuxta et abuttant super foreram predictam	*j acr* Rad.
Inter.ij.balc'		j.selio Rogeri de harleston quondam Ricardi Tableter iuxta et abuttat super foreram predictam	*di acr* Egid.
mo		ij.seliones Stephani Moriz senioris quondam patris sui iuxta et abuttat super foreram predictam	*iij rod* Egid.

THE WEST FIELDS OF CAMBRIDGE

tylers

sine balca

hospitalis
$ cχ

nota

✶

⎰ j.selio Johanis baldewyn quondam Aubre de Stowe iuxta et abuttat
⎪ super foreram predictam *hospitalis*   *j rod* Egid.
⎪ j.selio Willelmi Aleyn de bokesworth iuxta et est longior aliis
⎪ selionibus antedictis ad capud suum orientale ac etiam longior aliis
⎪ selionibus sequentibus ad capud suum occidentale et in parte forera
⎨ ad utrumque capud et abuttat ad capud orientale super pratum
⎪ mortimer   *di acr* Par.
⎪ j.selio (Ricardi martyn *deleted*) Johanis baldewyn quondam Andre Stowe
⎪ iuxta   *j rod* Egid.
⎪ j.selio Collegii predicti quondam henrici Tanglemere iuxta   *di acr (cl)* Rad.
⎩ j.selio Ricardi Martyn quondam Rogeri Russell iuxta *rede*   *di acr* Rad.
ij.seliones Johanis de Cotton quondam Thome de Comberton iuxta
et sunt ultimi et borialissimi illius quarentene et ultimus selio est
forera   *di acr* Botul'

vj.seliones hospitalis predicti iacentes transversi ad capud orientale.
iij.selionum hospitalis predicti ultimo dictorum ad tale signum. ✶ et
iacent in longitudine iuxta pratum domini Roberti Mortimer antedictum
et abuttant ad capud eorum australe super le Carmedole predictam
et ad capud eorum boriale super foreram Willelmi Aleyn de
bokesworth ultimo dictam quorum occidentalis selio est forera   [– –] Egid.
      Et sic finitur le Carme Feld

Summa selionum istius campi decimalium ecclesie botulphi xxviij
      Summa selionum istius quaterni liiij   summa totalis lxxx et x
   summa Johanis et Thome de Cant' iij^{xx} xj acra j rod
   summa Galfridi Seman   *iiij^{xx} xv acra iij rod et di*

b   xviij

APPENDIX A

I *Topographical notes based on the Terrier of* 1566 *in St. John's College* (*xxxi.* 24). *Spelling is modernised. A copy of this, made by Mr. Stevens in* 1800 *from a copy of the Terrier of* 1572, *is in Jesus College* [*EST.* 4.7].

Be it had in mind that there is many and divers names, terms and buttings in the four fields above rehearsed, the which perfectly known, a man may soon know every furlong in every field, the which begins in Greathowfield.

Greathoway begineth at the great pits without the Castle, before the Stumpe Cross on the west side, and goeth through to Greathow Hill, and so down the hills on the south side of the Conduit Head into a green way that goeth to Moor Barns called Bradrush.

Stowpencrouchwaye beside Chalkwell at the corner of St. John's Close, and goeth northwest unto St. Neot's way at the side Hewnell Close, now St. John's Close, and it parts Middlefield from Greathowfield, but that furlong is sown with Greathowfield.

Milneway or Milne Path begins on the north side of Greathowhill a little above the Conduit Head towards the east, and so straight to a little furlong that lieth next Girtonfield that begins with iij lands, sometime vj lands of Thomas Atchurch of House.

Braderushe Green begins a little beneath the Conduit Head on the west side, and so goeth to a broad balk that parts Cambridge Field from Girton Field.

Wilwayeditch begins at the Conduit Head, and at the beginning it is called Semonds Ditch, and it cometh running down towards the west side of Howe Nel Close called Fresh Croft, and so goes down to a way between ij little hills beyond the furlong called the Erber, and runneth into a green little meadow against the parcel of willows called Cotton Way.

The cross mound in St. Neot's Way is now but a little hill, and it is in the midway of St. Neot's almost at Morebarnes.

Cotton Way begineth at the parcel of willows and goeth straight to Cotton.

Endlesse Way begins behind all the furlongs that be near St. Neot's Way next Cotton Field, and at the beginning is many leys, and in winter full of water, and it goeth straight to Cambridgeward and endeth beside a furlong that lieth against Hedwell Close, and the most part of the furlong is sown with Carmefield.

Barton Way begineth at Chalkwell on the north side of St. John's Close, and goeth overwhart St. Neot's Way, and so by Hamole Close to Sheepcott Willows and so over a greenway to Aldermanhill on the north, for the other way towards to the south is called Custes Balk, and it goeth still over Edwin Ditch, and so to Barton Cross that standeth on the farther side of all Cambridge Fields at the end of Clintway. Alldermans is over against Shepcote Willows.

Dedall Furlong lieth under Cotton Field at the beginning against a long row of willows, and they be on the west side of Edwin Ditch, or else lieth on length, and be iij furlongs lying together towards Cambridge.

Clintway begins next Cotton hills at a great balk that parts Cambridge Fields and Cotton Fields above on a hill, and it goeth on the far side of all Cambridge Fields, and parts Cambridge Fields from Barton Fields, and it endeth at a cross of tree that standeth in Barton Way by a great row of willows that grow beside Portbridge.

Edwindich begins at Portbridge toward Barton, and it comes down between Littlefield and Carmefield and parts them, and so goeth down to a Close called Martinhall Close, and it runneth into the common river. (This is in fact the Bin Brook).

Custsbalk, otherwise called Custsway, begins over against Shepcote Willows upon Alderman's Hill of the south side, and it goeth over Edwindiche, and over a little way that comes out from a little lane from Newnam, and so to Old Newnham Way where there groweth divers willows, and so to Granciter Strait.

Olde Newneham Way begins a good (sic) without Newenham to Granciter ward at an old tree cross, and it goeth between Granciter Fields and Cambridge Fields till you come toward Portebridge toward Barton.

Longe Balk, otherwise called Milnepath Way begins at the Welhed beside Binebrook and it goeth toward Granciter over a little way that comes out of Newenham to Barton Way, and so it goeth into Granciter Fields.

Froshlake Way begins with a little way lane that cometh out of Newenham, and so goeth into the fields to Portebridge.

Cottonway begins without Newenham on the west a side of Longe Balk, and so goeth to Bynbroke and over Bynbroke through the Little Field.

Long Meadow is without the King's College, and sometime all the gardens and closes that the King's College hath now without their great bridge toward Cotton Way called Long Meadowe sometime before the King's College had it purchased of the commonalty of Cambridge.

II Notes, mainly topographical, from the later version of the Terrier, 1617, in St. John's College (xxxi.29). Spelling is modernised.

A perfect Terrar made out of Blackamore, of all the lands lying in the Fields of Cambridge on the West part of Cam, divided into three several seasons or times of ploughing. For one season or part of the field lieth fallow every third year, and so they are all set down in this Terrar as they are severally and yearly laid out, 1617.

Note that in this Terrar nothing is altered from Blackamore which doth concern either the contents of the land (which appeareth to be unequal) or the setting out or appointment of the tithe, but therein they do agree truly. As for the fields and furlonds they are somewhat altered; for they are better known by their new names and more plainly described than they have been heretofore for the more easy and ready finding out of every furlond in every field as followeth.

And first for the first season which doth lie in two parts the one distant from the other half a mile at the least, yet they are ploughed and sown together yearly, and alter not.

The first part of this season doth contain all Carmefield (which is better known by the name of College Field), Littlefield, and a part of Middlefield, viz. the 21, 22, & 23 furlonds and no more.

Description of Carme Field

This Carmefield (or College-field) is that which lieth next the town, betwixt Longegreene on the East and Binbrook on the West. There are in this field certain crofts, which be every year land and never left fallow, but at the will of the owner. These crofts are contained in the 1, 2 and part of the 5th furlonds, and all the 6th furlond croftland and no more.

Description of Binbrook and of Long Green

Binbrook is a drain that runneth from New Close in Grantchester and doth divide the College Field from Little Field till it cometh to St. John's new brick wall, and from

APPENDIX A

thence it runneth into the common river not far from the great bridge. Long Green before mentioned was wont to lie open and common from the Small-bridges unto Binbrook, where it so runneth into the common river as is aforesaid.

Description of Littlefield

Littlefield doth lie betwixt Binbrook on the East and Bartonway on the West. And it doth begin at Clintway, which leadeth to Portebridge, and doth end next St. Neot's Way at the town's end.

Description of Clintway

This Clintway doth divide part of Coton Field from that part of Middlefield which is sown with Carmefield and no more. And so goeth down to Portebridge, and from thence to Newnham, but at the bridge it leaveth the name of Clintway.

Note to Furlong 9 of Carmefield:

Note that against this furlond there is no common, but all several, for the selions do abbut upon Binbrook.

Further note to Furlong 11:

Note that all the lands do abbut Binbrook which do lie on the east part thereof in this field, which doth manifestly show the lands' ends not to be common at all times, but proper to the landholders.

Description of Custosbalk (p.12):

This Custosbalk doth begin here, and goeth down to Binbrook and so into Carmefield as you shall find it in the 19th furlond there, and so straight on towards Grantchester Field.
p.14: Now followeth part of Middlefield, even so much as is joined to these two former fields, which altogether, do make the first part of the first season, viz. the 21, 22, & 23 furlonds.

p.15: Thus endeth the first part of the first season contained in Carmefield, Littlefield, and these three furlonds of Middlefield viz. 21, 22 & 23 furlonds, which are always sown together, and lie fallow together.
Note that these three furlonds do lie just against the furthest part of Littlefield, as you see they do abbut on Clintway not far from Portebridge; and there is nothing that doth part these three furlonds from Littlefield but only Bartonway.
Note also that Clintway which leadeth from Coton and Bartonway and Edwinditch do compass these three furlonds and none other, and they do lie almost in a triangle.

Description of Edwinsditch:

This Edwins ditch hath his first beginning at Coton, and cometh down to St. Katherine's Willows, and from thence down by the first and second furlonds of Deddale, which are the 18 & 19th furlonds of Middlefield, and all the length of these two furlonds it doth part Coton field and Cambridge field, and at the East end of this 2th furlond of Deddale it turneth a little into Cambridge field even at the beginning of 21 furlond of Middlefield before mentioned at a great piece of 20 acres of the Clerks of Merton, and from thence leaving all the former three furlonds, viz. the 21, 22 & 23th furlonds of Middlefield all towards the South, it runneth over Bartonway, and so into Littlefield as you shall find it there mentioned in the third and fourth furlond, and from thence into Binbrook.

Description of Longbalk:

Longbalk is often mentioned in Carmefield, it doth begin next to Binbrook in the 10th furlond betwixt three lands of Clarehall on the one side and two lands of Thomas Morris on the other side, which two pieces do abbut upon Binbrook. It goeth also through the 12th furlond and so directly towards Grantchester as you shall find it plainly in these 2 furlonds, viz. the 10th and 12th.

p.16: This part of the furlond (8) beginneth at 2 selions of Benet College on which two selions there was erected a dwelling house in 1603.

p. 17: There is mention in Blackamore of a cross and of a little hill standing in St. Neot's Way at the end of these furlonds but there is none such now, neither hath been of late that can be remembered.

p. 18: 1 selion of Roger Harleston, and it is the last and most east next unto St. Neot's Way, and it lies unploughed. (Fur. 11)

p. 20: 14 selions of the Prior of Barnwell and they be called Duckdole, and are all sward or grassground.

8 selions of Benet College. These 8 Selions do lie at the upper end of one part of those 11 mentioned at the latter end of the 8th furlond, and they are ploughed through together as though it lay all in one furlond.

Note also that these 8 selions are parted from Duckdole before mentioned with a ditch and quickset, the bank cast towards Duckdole yet on the College ground.

p. 22: 4 selions of the Hospital; on the south head of the last selion sometime there stood a sheepcote. These selions be the most west of this furlond and the tithes of them do belong to the Hospital, and in old time, as it is mentioned in Blackamore, this last selion was a sanctuary.

p. 23: Note there are two selions which do lie crosse of part of this furlond of the monastery of Denney and next along to Willow Ditch, which are not ploughed, and they seem to be a little furlond by themselves (Next Porters Dole).

p. 24: 8th furlong in Middlefield:
Thus far this furlond lieth with this field and no further. This furlond is described in the second part of the first season unto which part of this furlond doth belong; and the other part doth appertain unto this season, viz. from the end of Endlessway, all the other part which is homeward towards Cambridge.

p. 27: From hence do follow Cupids or Cupis crofts which are thought to be every year land, but of late they have been added to the second part of the first season, and they do all lie next unto Coton and do abbut on Endlessway at the one end, and Sheepcoteway or Coton Way at the other end.

...... which is most towards the west, notwithstanding Blackamore doth call that the north end of the furlond.

p. 30: (19th furlong): 2 long selions lying next Edwinsditch and a broad butt. And as they do lie along by it so do they also abbut upon the said ditch; for at the end of these two selions the ditch cometh into Cambridge field, and doth part the two fields and no further, and all the furlonds which do lie on the west part of this ditch from henceforth do lie with Carmefield and Littlefield and do belong to the first part of the first season.

APPENDIX A

p. 32: (Middlefield, the third and last season): This season doth contain the first furlond of Middlefield, and all Greathowfield: St. Neot's Way lieth on the south-west part of it through the whole field.

Description of Stopendcrouchway: (p.33):

Stopendcrouchway before mentioned, and after to be mentioned again in Greathowfield doth begin at the hills not far from the watering pond, and goeth down with a crooked winding even unto St. Neot's Way, and it doth divide Greathowfield (so far as it goeth) from Middlefield.

p. 34: (Furlong 4): This furlond doth only contain 10 selions of the Hospital with an headland; it lieth betwixt Greathoway on the one part, and Stoupendcrouchway on the other part, and it doth abbut at the south head upon the Claypits or Chalkpitts of St. John's opposite to Harr Hill, and they be one furlond by themselves.

p. 36: (End of furlong 6): Thus far lieth Carmefield, but the bounds of Cambridge lieth further into the closes as doth appear in Blackamore.

p. 38: (Furlong 8): 1 selion of the Clerks of Martin (sic); this also should abbut upon Greathowpath, and not go over the same right against How Hill, as Blackamore saith, but it is now ploughed much further.

p. 40: (Furlong 9): Note that these 4 last single selions are now made claypitts and not ploughed.

p. 43: (Furlong 14): 1 selion of Thomas Attechurch of Howse the first and most south; Blackamore saith is it an headland, but it is now altered and is not so by reason much gravel hath been digged and made pits, and now ploughed again otherwise than it hath been at the beginning.

p. 48: (Furlong 23): . . . the high cross which is now but a little hill in St. Neot's Way.

p. 49: (Furlong 26): This furlond lieth cross of the last furlond and doth abbut upon St. Neot's Way with the West head, and upon Maddeleye Moore with the East head. This furlond is all inclosed in Maddelye Close.

APPENDIX B

Mortimer Charters in Gonville & Caius College
Charter of William Mortimer to Adam Dunning (G&CC Box XIII I (A, B))

Notum sit omnibus presentibus et futuris Quod Ego Willelmus de Mortuo mari dedi et concessi et Hac presenti Carta mea confirmavi Ade filio/
Eustachii totam terram meam Integre sine aliquo retinemento excepto mesagio/ quod fuit Beniamini quam vero terram tenui de Comite David scilicet in Cante/
brig' Neuham et Bernewell. In villa et extra In dominicis serviciis homagiis Molendino Redditibus cum quatuor Libris de tercio denario Comitis/
David de villa de Cantebr' cum omnibus pertinenciis et escaetis que eidem terre possunt evenire In pratis pascuis et pasturis Tenendam et Habendam de me/
et heredibus meis eidem Ade et heredibus suis Libere quiete Honorifice In feodo et hereditate Reddendo inde annuatim mihi et heredibus meis triginta/
duas marcas ad tres anni terminos scilicet ad festum sancte Etheldrede decem marcas et ad mediam quadragesimam decem marcas et ad festum Sancte
Margarete duodecim marcas Et novem solidos et quatuor denarios pro me et heredibus meis ad Haggablum Ballivis domini Regis de Cantebr' pro/
omnibus serviciis consuetudinibus et exactionibus Et Ego dictus Willelmus et heredes mei warentizabimus totam predictam terram cum omnibus pertinenciis/
et escaetis predicto Ade et heredibus suis Contra omnes homines et feminas et versus dominum Regem de Scutagio et versus alios capitales dominos
de omnibus serviciis Et Sciendum quod Ego Willelmus neque heredes mei possumus neque debemus Exigere aliquod relevium vel Wardam de predicto Ada/
vel heredibus suis propter magnitudinem firme sue Pro Hac autem Donacione Concessione Carte Confirmacione et Sigilli apposicione/
dedit michi predictus Adam quadraginta marcas In Gersuma. Hiis Testibus. Domino Petro Constabulario de Menton'. Domino
Roberto de Beche. Domino Math' de Morley. Herveio fil' Eustachii. Roberto Seman. Herveio fil' Mart'. Gaufr' fil' Rad'. Mich'/
fil' Herv'. Johanne de Sexton. Johanne de Scalariis. Thom' fil' Joachim'. Reg' Doy. Will Pilat' et Multis Aliis.

Dating
William de Mortimer, s. of Sir Robert de Mortimer—fl. 1200.
Earl David of Scotland (brother of William the Lion), granted the Honour of Huntingdon with the County of Cambridge in 1184. Died in 1219. His wife was d. of Ranulf, Earl of Chester.
Adam Dunning, succeeded by Leonius by 1243 but living 1239–40.
Robert de Beche—fl. 1216–36.
Matth. de Morley, perhaps s. of the earlier —— de Merlay—fl. c. 1200.
Herv. fil. Eustach'—fl. 1195–c. 1231–2.
Herv. fil. Martin'—fl. 1220.
John de Sexton=(?) Jeremy de Caxton, Sheriff, 1232 & 1234–6.
John de Scalers—fl. 1225–45.
Thos. fil Joachim=Thomas Tuillet.
Ref. mesagium quod fuit Beniamini=the Tolbooth, formerly of the Jew, Benjamin of Cambridge.

APPENDIX B

Charter of Richard Dunning to Robert his Son and Wife Joan, 3 Edward II
(G&CC Box XIII IB)

Sciant presentes et futuri quod Ego Ricardus Dunnyng de Cantebrig' dedi concessi et hac presenti carta mea confirmavi Roberto Dunnyng filio meo et Johanne uxori eius omnes terras meas
et tenementa mea de feodo domini Constantini le Mortymer videlicet unum pratum ad fratres de monte carmeli ultra aquam ad Newenham. Et ibidem unum molendinum aquaticum
cum quodam prato adiacente Iuxta le Dam. Et septem acras terre cum pertinenciis in Bynnebroc' et sex acras terre cum pertinenciis in Spiteldole. Et octo acras terre cum pertinenciis Iuxta
terram sancte Radegunde que abuttant super London' Waye. Et decem acras terre cum pertinenciis in Swynecroft. Et decem acras terre cum pertinenciis ad Pyschewellewaye. Et octo acras terre cum pertinenciis apud Hynton Waye. Et decem acras terre cum pertinenciis ad Brademere. Et tresdecim acras terre cum pertinenciis in Bernewellefeld. Et septem acras terre cum
pertinenciis ad Hynton moer. In Portefeld apud Aldyrmanhyl viginti acras terre cum pertinenciis. Et sex acras terre. In Le Clay. Et tres acras terre ad Stonydale. Et tres
acras terre apud Bertonwaye. Et tres acras terre retro Mertones que abuttant super Le Broc. Et quatuor acras terre ad Chalkpyttes. Et tres acras terre ad Chalkwelle. Et tres acras et dimidium et unam rodam terre apud le Stoupande Crouche cum libero Ovili et libero Tauro et libero Apro. In Campis de Bernewelle et in Portefeld. Preterea sex marcas argenti annui Redditus annuatim ad percipiendum de villa Cantebrig' ad duos anni terminos videlicet ad festum sancti Michaelis Tres marcas et ad Pascha tres
marcas. Scilicet de heredibus Ricardi Bateman octodecim denarios ad predictos terminos. De domino Waltero vicario ecclesie sancti Clementis tres solidos ad eosdem terminos. De heredibus Roberti
Steresman per annum quatuor solidos. De Reginaldo de Cumbirton per annum sex denarios. De Le Kaye apud Dame Nycholeshythe per annum octo solidos. De Laurenc' Seeman per annum
viginti duos denarios. De Alano de Welle per annum duos denarios. De heredibus Johannis Edes per annum duos denarios. De Willelmo Pyttok quatuor denarios. De Matilde de Welde per annum
duos denarios et obelum. De Johanne Porthors obelum per annum. De Henrico Matfrey per annum duos solidos. De Willemo de Bekyswelle duos solidos. De uxor Johannis Bernard de Bernewelle
decem denarios et quatuor capones. De magistro Thome Tuylet sex denarios. De Willelmo Tuylet per annum decem denarios. De Priorissa Sancte Radegunde octodecim denarios. De
Capella sancti Edmundi quatuor denarios. De Radulfo de Cumberton duos denarios. De Rogero Audre quinque denarios De Johanne de Schelford' sex denarios. De Roberto filio Henrici Tuylet septem denarios.
De magistro Ade Elyot duos denarios. De Waltero Sparewe de le Coten duos solidos et quatuor capones. De Johanne de Marbelthorpe quatuor denarios. De Johanne Saladyn de Bernewelle duos solidos et quatuor Capones. De Sylvestro de eadem quinque denarios. De Laurenc' Dixy tres denarios. De uxore quondam Johannis Dunnyng per annum quinque solidos et sex denarios. De Roberto Sabyn de Gran-

teseth octo denarios et duos capones. De Johanna de Oxenford duos solidos et duos capones. De Henrico Pestour et Galfrido de Cumberton sexdecim denarios. De Herveo Pilat tres solidos.
De Johanne de Trumpiton unum denarium. De Johanne le Comber duos solidos et duos capones. De Johanne Martyn duodecim denarios. De Julian' le Spicer unum denarium. De Johanne fil' Hugonis sexdecim denarios.
De Johanne Golderyng Triginta et duos denarios. De Andrea de Hychen duodecim denarios. De Petro de Bermyngham quatuor denarios. De Symon de Bradele sexdecim denarios. De Hospitali sancti Johannis
sexdecim denarios. De Waltero de Berkyng sexdecim denarios. De Sayero molendinario quatuor denarios duos capones. De Malina Mydewyf unum denarium. De Roberto Sabyn octo denarios. De Matilda Morys
quinque denarios. De Johanne le Charner duos solidos et duos capones. De Milone de Trumpyton duos solidos et duos capones. De Reginaldo de Costeseye duodecim denarios. De Ricardo Wetherysfeld sex denarios.
De Avycia Martyn sex denarios. De Stephano at ye kilne tres denarios. De Willelmo le Roper tres denarios. De Joachim unum denarium et unum par Cyrothecar'. De Symone de Stokton unum capon.
De Bricio le Mounford octo solidos. De Johanne le Eyre novem solidos. De Jordana unum denarium. Et similiter servicia omnium tenencium meorum in Cantebrig' Bernewell Newenham
et in Howes cum omnibus suis pertinenciis. Habendum et Tenendum omnes predictas terras et omnia predicta tenementa cum pratis pascuis et pasturis una cum predictis libertatibus et omnibus annuis redditibus prenominatis de Capitalibus dominis feodi per servicia inde debita et consueta predictis Roberto et Johanne uxori eius et eorum heredibus de corporibus legitime procreatis libere quiete Integre
bene et in pace sine alicquo contradiccione Inperpetuum. Et ego predictus Ricardus Dunnyng et heredes mei
omnes predictas Terras et omnia predicta tenementa cum omnibus Redditibus. libertatibus et aliis suis per-
tenenciis nominatis et non nominatis predictis Roberto et Johanne uxori eius et eorum heredibus de corporibus legitime procreatis contra omnes gentes Warantizabimus Inperpetuum. In cuius rei testi-
monium huic presenti Carte sigillum meum apposui. Hiis testibus Symone de Repham tunc maior' Cantebrig'. Henrico de Toft. Willelmo Carbonel. Symone de Bradele tunc ballivis
Ricardo Laurenc' Michaeli Pylat. Laurenc' Seman. Johanne de Mabilthorpe. Johanne Goggyng. Thome Audre et aliis. Datum apud Cantebrig' die dominica proxima post festum Beate Katerine/virginis. Anno Regni Regis Edwardi filii Regis Edwardi Tercio.

(29 November 1309)

Seal of Richard Dunning, triangular with a device of a snipe-like bird (? a dunlin).

APPENDIX C

A Note on Trees in the West Fields

Soon after the laying-out of the new Trinity Fellows' Garden, in land acquired at enclosure, there was a joke current in the University;
Q. What sort of prospect has a Fellow of Trinity?
A. A long, narrow prospect, with a Church at the end of it.[1]
The Church was, of course, Coton Church, whose spire was visible from the newly-planted avenue across the still open expanse of arable West Field, for the new hedges took some time to grow up. Except that the avenues and groves laid out on the College "Backs" had crept a little further westward, the general appearance of the town from West Fields, viewed across almost tree-less arable, must have differed little from the open Prospect of Cambridge from the West, made familiar by Loggan. But there were a few trees in the West Fields, some even mentioned in the Terrier, and some which came with earlier enclosure. Although all indigenous woodland had been cleared in the interests of corn-growing long before the 14th century, there is enough evidence of limited tree-growing to be of interest to the historian and to the botanist.

Willows. The only species of tree mentioned regularly in the Terrier is the willow. We find willows growing along Willowes Ditch, and the additional notes describe the position of St. John's sheepcote with reference to the "ixth' or xth welowe on the north side of berton weye." (fol. 14v.). The St. John's and Jesus terriers tell us of St. Catherine's willows on the Coton boundary, and of the large clump of willows growing at Portbridge. Willows were encouraged in the few remaining damp green places of the West Fields, even though there was no special place set aside for them, such as on Empty (Impty) Common on the other side of the river. The withies were extensively used for basket making, and the basket makers appear (from 16th century deeds in C.C.C.) to have congregated chiefly in Bridge Street, opposite Sidney Sussex College.

Elms. There is little doubt that elms were the fashionable tree for planting, on field boundaries and College walks alike, from the seventeenth to the nineteenth centuries. Expert botanists can ascribe a date to an elm avenue by the species of elm alone, different kinds being favoured at different periods. So it is not surprising to find the close at Edwin's Ditch, marked "Merton College Esq." (!) in the 1789 map, planted with an elm hedge that can still be seen. Elms again mark the boundary between between furlong 17 and the top of Dedale. Elms may sometimes have grown on balks, for there are the remains of ancient elms on Lammas land, in places corresponding to the balks shown on the 1805 pre-enclosure map and to the edge of furlong 1, which were certainly not put there by the Parks Department of the Corporation.

It is also possible that a few elms, and perhaps other trees, survived on the small greens. There is a very old elm on the site of Kynch's mead. Loggan's print also shows a group of trees at the junction of the Barton Way but one would hestitate to pronounce on their species, for though they might be the shape of elms they are about the size of the hay-cart.

1 Contributed by Professor Dickins.

Oaks. The indigenous woodland which was cleared away from the far west of the West Fields was most probably oak woodland, such as survives in Madingley Wood. Oaks grow well in this area, and the surrounding of the Moorbarns Close by a band of oak woodland has produced some very fine trees. That the woodland here is pre-Enclosure is certain from the fact that it follows the shapes of the original Furlong 24, not of the later boundary. Without felling the trees it is not possible to pronounce exactly on their age, but 200 years would seem a moderate estimate.

A single oak with a girth suggestive of greater antiquity than Enclosure stands at the North West corner of Caius cricket field, formerly the boundary of furlongs 2 and 6 of Carmefield.

Mixed Scrub and Hedgerow. Following the now accepted rule of the (Cambridgeshire) botanists that the greater the number of species in a hedge, the greater its antiquity, we may identify a few pieces of remaining ancient hedgerow. The best of these is by Edwin's ditch, just where it turns inward from the Cambridge-Coton boundary. The bank by the ditch here carries enough species to be ascribed to the later Middle Ages. Another part rich in species is the line of the Binn Brook itself, as it crosses the site of Little Field, and is joined by Edwin's Ditch. There are interesting patches of mixed scrub and trees at Conduit Head, over former gravel pits, and below the Howes, but the proximity of houses and gardens makes precise dating impossible.

Walnuts. We have not found any walnut trees from before Enclosure, but the old walnut trees at the end of the former garden of St. John's Farm must surely have been planted as soon as the new farm was built, following the common custom of Cambridgeshire farms. Self-sown walnuts appear among the mixed shrubbery by the Binn Brook, mentioned above.

Hawthorn, or "quicks", were the standard post-Enclosure hedging, but must also have been usual for early enclosure, for they are specifically mentioned as surrounding Dukdole.

APPENDIX D

Place and Field Names in the Cambridge West Fields from Corpus Terrier (now Add. 2601)

Folio				
		GRITHOWFELD or GRYTHOWEFELD Grythowe, Gretho, Grethoe,		later corrupted to Great Howe (18th century). Mound or tumulus at
		(13th century)		the south end of the gravel ridge which gives its name to the Field.
		Grythoweweye, Grithoweye		leads to the Grithow from Castle Hill.
		Grithowepath		leads to the same from the Madingley Road.
1		Asshwykston		The older cross in the Huntingdon Road at Castle End to the south of the better-known 16th century cross (see *Bernewell* p. 168.)
		Stoupendecruche		A bent cross in the Madingley Road, giving its name to a way from thence to Castle Hill.
		le Cleypittes		Just outside the Upper town on the side of the Huntingdon Road. Still visible as lowered ground-level with waterlogged hole.
		Hor Hil		Later Mount Pleasant, just outside jurisdiction of borough officials. Variously mistranscribed as Horse Hill and Hare Hill. A similar name to S. of town off the Trumpington Road.
	Flg.			
2	3	Le Sal, Sale Piece		A five-acre piece on north side of Castle bounds, anciently held by the Dunning family and perhaps the site of their family home (i.e. sael=hall).
		le Sponyaker		Possibly wet ground (?)
		le Blakaker		Probably from the appearance of the soil.
		litilmer' or Litilmor'		A small pasture, hardly more than a widening of Grithowweye, but probably a relic of a larger pasture—see Priorsdole.
	4	Sengilakir		single acre.
3		Le Howes		Hamlet at the crest of the gravel ridge on the Cambridge boundary. Ancient settlement and early enclosure. Early called Howescroftesande Howeshedge in 1789 map.
3v.		Howescroft		
	5	Milneweye or Mulleweie		leading from the Grithowe to a mill beyond the Girton boundary (?site of later French's mill).
4	5	Le greneplat		Rough pasture ground at Grithowe end of Millweye, early used for gravel digging and allotted to the poor at Enclosure. Cf. use in Derbyshire for the green area adjacent to a farmhouse.
		Le Grenesheld (early 13th century)		

THE WEST FIELDS OF CAMBRIDGE

Folio	Flg.		
5v.	6	Priouresdole	Late assart, formerly pasture, cut by Grithowweye. Allotted to St. John's Hospital and which "Prior" is uncertain, possibly an exchange with Barnwell.
		le Cuttedrode	Strip ploughed in two sections, possibly because of contour.
		Tunmannisaker	Townsman's acre, perhaps originally connected with some town office but in hands of Morys family by 14th century.
6r.	7	Brembilfurlong	An unpromising area, see Nakededole.
	7/8	Dukmere or Wlwyesmere	Here gravel gives way to clay and a small damp patch, probably late assart, is separately ploughed as Furlong No. 8 "parva quarentena".
		Nakededole	Part of Bramblefurlong (above).
7v.	9	le Gyldenaker	An ancient possession of Corpus Christi College, via The Gilds of St. Mary and Corpus Christi. Another of this name in Middlefield.
8v.	12	Le Cunduyt or cundut	The Franciscan conduithead at the S.W. of the Grithowe.
		Peperdole	On south slope of Grithow adjoining Seman's ditch (see below) and formerly belonging to Geoffrey Seman in Terrier. Early enclosed (Peper=pebble).
		Semannis dich	Carries off excess water from Conduit Head area to south side of St. Neots Road. Possibly dug by Seman, head of a well-known Cambridge family, found early in Cambridge and Newnham; owned much land in 13th century including all the area adjacent to this ditch.
8v.	13	Kynchmade	Streams flowing westwards, with adjacent pasture strips, carry water from the springline of the Grithow to the Washpit Brook; the upper pasture is Kynchmade (or mead) and contains ancient elms.
9v.		Braderussh	The lower stream is Braderasshe or Broadrushe. Course of both preserved by exisitng field ditches.
9r		Grittonfeld	Fields of Girton; Grit-tun.
	15	Le Mordole	Adjacent to Madingley Moor and the S.W. boundary of Girton parish. A late assart. Le Morhyl also found (13th century).
10v.	18	Le Goredaker	A selion wider at one end.
12r.	22	Le chekker	A reference to the building of that name in Barnwell Priory to which account was made. Most of this and the adjacent furlong belonged to the Prior of Barnwell.

APPENDIX D

Folio	Fig.	Name	Description
	24	Le Morbernes or Moor Barns	An extension of the Cambridge West Fields into Madingley Moor. Early enclosed and for practical purposes treated as part of the Madingley estate from the 16th–17th centuries.

MIDDELFELD

A composite field. The name first applied to the area of Le Cley (early 13th century) Portefeld the older name for the greater part of this field (in 13th century).

Folio	Fig.	Name	Description
1	(23)	Muscroftfurlong / Muscroft }	The earliest enclosed croft near Castle Hill, possibly same as Chalkwell Close (qv.) ('Mus' or 'moss'=boggy.)
		Saint Iones bernys / St. John's Barns	The headquarters of the St. John's farm for the Cambridge field (also called The Grange later—see Le Longebalk).
13r		Bertonweye	former Roman road entering Cambridge (upper town) from the S.W. An important green way prior to the making of the turnpike (modern Barton Road) and a Field boundary.
13v		Chalkwelle / Chalkwelledole / Chalkwellcroft	A spring at the S.W. corner of the upper town, later 'Drakes Spring'. Adjoining dole allotted to Mortimer. Croft allotted to St. John's Hospital by Baldwin Blangernon (St. John's Coll. Muniments).
14r	2	Frosshiscroft	A damp (froggy) spot on south side of Madingley Road opposite How Hill, early enclosed. Later corrupted to Freshcroft.
	2	Wlwesdich and Wilwys Close (=Frosshiscroft)	Ditch or stream fringed by willows, draining towards the Binn Brook.
	2 (24)	Seyntgilis aker	Owned by St. Giles' church.
14v	3 (25)	Hunnell Cross	In Barton Way at junction by St. John's Sheepcote (16th century name).
		Hunnell's Close	(Also 16th century.) Another name for Freshcroft.
		(St. John's) Shepcot / Shepecotte	"At the 9th or 10th willow in Barton Way" by Hunnell's Cross (above). "Le bercarye" (15th century).
15r		Smalemade	A very narrow green pasture by St. John's Sheepcote, probably relic of a much larger pasture, watered by Willowes Ditch. (Smal=narrow)
	4 (26)	Cotesweye	Also called Sheepcoteways, from the Sheepcote towards Coton. Replaced after Enclosure by the Coton footpath.
	5 (27) 6	The Heerne voc' le erbeer	In the corner formed by a bend in Willowes Ditch (Hearn=corner, erbeer=harbour).

THE WEST FIELDS OF CAMBRIDGE

Folio	Flg.		
16r	(28) (29)	Porthorsdole	A two-acre strip alongside Willowsditch, named from John Porthors, Cambridge burgess and landowner of 13th century.
8a	(30)	Brunneforthdole	A 6-acre patch adjacent to the above dole, where it must have been possible to ford the ditch. (See Le Brunneford, Maitland T. & B. p. 172.)
8b		Le Cley–Le longe furlong in le Cley	an ancient balk separates this area of heavy clay land from the more ancient arable of Furlongs 1–8a.
17r.	9	The furlong beyond High Cross (16th century) "ultra altam crucem"	Furlongs 9 and 10 lie to the West of the ancient cross in the Madingley Road which marked the former boundary and are assarts towards the Cotes. Strips here largely acquired by Coton-based owners and by 16th century farmed from Coton.
		Godyinesrode	In 13th century charter (C.C.C.VII.14), probably in this furlong but name lost by 14th century terrier. (Cf. 31r Godynys Buttes.)
17v	10	Sheriffesdole	At West end of this furlong. 16th century name and almost certainly refers to name of the holder, not the office.
18v	14	Endelesseweye	Endlesweye Path—green way from Coton ending abruptly in mid-field—referred to by Maitland (T. & B. p. 123).
19r		dukdole (16th century)	a damp area soon reverting to pasture and enclosed, belonging to Barnwell.
		dukpytte (16th century)	reference by Maitland (T. & B. p. 123) next to the above, belonging to C.C.C.
20r		le Gildenaker	Another of this name in Grithowfeld (see above).
21r	15	Thorpiscrott	A grove, 3 or 4 strips wide, alongside Cotes Weye towards the West end. Probably used as orchard or for loppings.
21v	16	Le Daleweye	On the Coton boundary separating Coton and Cambridge fields and leading to Deddale or Le Dale.
21v	16	Sparwescroft	The last selions of furlongs 16/17 at the Coton boundary—from a personal name as in Grithowfield (above).
		Blakaker	
22r		Karlokaker	Possibly from the arable weed Charlock, but a Richard Karlok appears in the Hundred Rolls.
	16	Aldermannehyl Aldermanishyll	Land rising to about 50 ft. beyond the Edwinbrook/Binnbrook junction. Possibly waste until a relatively late date, and allocated in large blocks (Mortimer's 42 acres). Earliest documentary evidence (C.C.C.VII 20) suggests use as field name antedates the emergence of the town official (c. 1288), if so, from Ealdorman and of pre-conquest origin.

APPENDIX D

Folio	Flg.		
			Possible connections with Edwin's ditch or defence arrangements for Cambridge in late Anglo-Saxon period.
22r	17	Ewynes dych and Edwyndich, Edwinbroc (13th century)	Tributary of Binn Brook coming from the Coton boundary and forming parish boundary in S.W. of the fields. Edwin not identified (see also Butebroc, below).
		Lykylt Grene (16th century)	Green patch where Edwin's ditch crosses Barton Weye, perhaps last remains of "longum pasture iuxta Edinbroc" (Maitland T. & B. p. 172).
23r	18	Le Deddale Dedemannesdale (1299)	"In ye lowe"—16th century, a stream valley (Edwin's ditch) in the far S.W. of the fields, formerly Le Dale (Merton 1606). (cf. Dedmannesweye in Grantchester).
		Erlesdole	15 selions held by Sir Bartholomew Burwash of Grantchester, part of the former lands of John Lacy, Earl of Lincoln (dec. 1232), on the far west boundary of Deddale.
		le Weie Balk	Carries the Coton path through the furlong of Deddale to join Le Daleway, above, and predates the ploughing-up of Deddale.
24r	21	Clyntweye	From the junction of Barton Way to the Clint (=hill) field of Coton parish. Forms the South boundary of the fields. Junction marked by Colys Crosse (16th century)
		Clynthavedene	also 'haueden' and 'Hevedlond' (C.C.C.XIV 1) means Clint headlands. Possibly between Qs. 21 & 23.(Maitland T. & B. p. 172).
		Goidzmedole or Goidesmedole	So named in 14th century by confusion with William Goldsmith, an adjacent owner, but in early documents (Merton 1606) appears as Godivesdole.
		Butebroc	(T & B p. 172) arable strips in the area beyond the brook. Name in contrast to Binbrook (q.v.)='super le broc'.
24v	23	Mundeslond	A personal name in all probability but no such owner found to date.
		Herwardaker—unplaced—	(from G. & C.C. XIII 2B)—obviously from Hereward, an unidentified owner.
24v		*LYTELFELD*	The name given in the 14th century to the arable furlongs lying between Barton Way and Binn Brook. Name used for at least furlongs 1–3 of this field in 13th century but at this time furlongs 7–10 probably part of Cambridge field. (Cart. of Sir J. de C. fol. 33v). By 18th century name had shifted to apply to the portions of Middlefield sown with Carmefield (8b–14)—see C.C.C. Map of 1789.

THE WEST FIELDS OF CAMBRIDGE

Folio	Flg.		
		CARMEFELD	—Used from the arrival of the Carmelites in Newnham in the mid-13th century, (see Carmedole below), later called College Field, by St. John's College.
28r		Eldenewenham or Neuham (13th century)	The name for Newnham village, particularly those parts most thickly settled in the Middle Ages, next to Newnham Mill. Sometimes used for adjacent arable (Qs. 1 & 2).
		Eldenewenham weie	Leading from the above towards Grantchester with a sharp-angled turn at the S.E. corner of the fields. Forming boundary with Grantchester from this corner to Portebrigge.
		Eldenewenham Croftes	Parts of Qs. 1, 2, 5 & 6, incorporated into the arable by 14c. and later detached into closes. Anciently applied to older crofts on both sides of Cambridge/Grantchester boundary. (C.C.C. XIV 12.)
	1	Mortymeres Dam	The mill-dam of Newnham mill, a pre-Conquest mill, formerly of Count Alan of Brittany from Ediva the Fair, who also had the Grantchester Mills. From c. 1200 to the Mortimer Manor at Newnham.
		Mortimeres Madewe	The meadow beside the cut to the Newnham mill.
		Melneyerde	A house with yard close to the Mill (above).
		Melnedampondes	Fishponds by the mill-dam owned by St. John's Hospital but with quit-rent to Mortimer manor.
	1	Le Lampeaker	Given to the Hospital by Hervey fitz Eustace to provide a lamp for the sick at night (Rot. Hund. II 359).
28v	2	Dufowshyll (16th century)	The dovehouse stood at the edge of Old or Elde-Newenhamweye (above).
		Ffrosshlake weye	Later corrupted to Frostlake way, from a small stream in what is now Newnham College, called the Frosshel- or Frog-lake. Now Maltinghouse Lane.
	3	Coten Path	Continuation of the above way towards Coton between furlongs (Qs. 3 & 7).
		Le Longebalk	Long balk running N–S cutting across furlongs. Of great antiquity, origin unknown. Later site of Grange Road.
29r		Custesbalk or Custesweye	Parallel and similar to the above, but further to the West. Both originally ended at Binn Brook but a continuation of Custesweye beyond the Brook joined Barton waye by 14th century. A not uncommon personal name but no Cambridge owner yet found bearing it.

154

APPENDIX D

Folio	Flg.		
29v	3	Shermannisrod	A personal name, though owner not identified.
		Bynnebrookes, Binbroc (12th & 13th century)	Anciently 'Le Brok', as in 13th century "super le Broc". The stream running into the Cam which originally took its name from the field knonw in 13th century as "In Binbroc" (see *Reaney* p. 44). Actually $2\frac{1}{2}$ acre piece wider at one end, ploughed in 6 selions and two gores.
30v	7	Goredaker	
31r		Godynys Buttes	At the end of Furlong 7, adjacent to the Binn Brook. To C.C.C. by 14th century but previously of John Godyn, chaplain.
32v	10	Gaggesaker	Probably from a personal name but previous owner not recorded.
32v		Le Longe Grene	Strip of common pasture nearest the river, later taken for 'The Backs'; the portion opposite Furlongs 10 and 12 later called King's College Green, and the access balk or headlands between them called King's College Way.
33r	13	Carmedole Furlong	From the Carmelite House based in Newnham for 40 years to 1292. After the departure of the Friars the house reverted to the Mortimer manor and was referred to as the Manor House of the Lord Robert Mortimer. This dole possibly ancient demesne of the same manor. The dole gave the name to the whole field from them id 13th century but was also called Spiteldole in the 13th century (G.& C.C. XIII).
	14	Carmedole	
34r	13	Barkeresaker	The Barker family lived in Newnham in 14th century. (Mortimer terriers and C.C.C. deeds).
	14	Whelpesaker	Outlying portion of Carmedole furlong—adjacent to pratum Mortimer (below), also to the Hospital.
		"Pratum Mortimer", or Butcher's Close (17th century)	A strip of meadow adjacent to the Carmedole and the manor. Early enclosed, with tree belt from c. 1800, possibly the only portion of this field which has never been ploughed.

APPENDIX E

Tables of Ownership in Five Areas of the West Fields

AREA I

Owner	Quondam owner	GRITHOW				MIDDLE			Total No. of Selions
		2	3	4 {East part	6	1	2 {East part	3 {except S. end	
Merton	(ex-Dunning)	(9 per se in Sale)	—	—	6	—	—	—	6 (+9)
Mortimer		—	—	—	—	4+3+2	—	—	9
St. John's Hospital		5	2+3+3 +1+2	2+1+3	—	5+4+5 +6+1	3+2	5+1+1 +4	59
Prior of Hunt.	(ex-de Troubelville)	6	—	—	2+1	—	—	—	9
Nuns of Beche		—	1	1	2	2	1+3+3+2	—	15
Stourbridge Chapel		—	—	2+3+1+1	2	—	—	—	7
Chantry in H. Sep.		—	—	2+1+1+1 +5+3	3	—	—	—	16
Corpus Christi College	G. Seman, T. de Cant'.	6 / 1	— / 2+1	1+1+1+2 / 1+4	2+1 / 1+2	2 / 2	1 / 2	— / 5	17 / 21
Moriz, Stephen (senior)	q. patris sui	—	3	3	4+7+1	—	3	—	21 ⎫
	q. J. de Berton	10	2+7	—	4+4	6	10	—	43 ⎬ 87
	q. J. Redhod	—	2	—	1	1	2	2	8 ⎪
Moriz, Thomas	others.	—	2+2 / 5+2+4	2 / 3+4+2	1 / 8	2 / —	3+1 / —	2 / —	15 / 28 ⎭
Bolle, Thomas	J. de Toft	1+1+1	—	1+1	4 (per se)	3	—	—	12
Brigham, Robt		—	—	8+2	—	—	7	—	17
Roger de Harleston	q. Wm. Warde	—	—	2+1	—	3	—	1+1+6	14
Others		—	—	1	—	3+1+4	—	3+1	13

APPENDIX E

AREA II

OWNER	GRITHOW				MIDDLE						Total
	21	22	23	24	9	10	11	12	13	14	
Prior of Barnwell	13+ 7	5+ 12+ 14	8+ 6	6+½ 3+ 8	2	1+1 +3	13+ 3	—	4+ 2+1	6+ 14½+ 2	125
Prior of Hunt.	—	—	—	9	2	9(11)	—	—	—	—	20
Hosp. of St. J.	—	1	—	—	1	1	—	—	—	—	3
Nuns of St. Rad.	—	—	—	—	—	—	—	—	8	3	11
St. Giles' Ch.	—	—	—	—	—	—	—	—	1	—	1
C.C.C. ex. G. Seman ex. T. de Cant.	— —	1 2	— 2	— 2	1+2+3 +9+3	— 2	5 —	10 butts —	4+4 butts +1+2 9	— 11	30+14 butts 28
Moriz, S.	—	3	—	—	—	1	—	—	2+2	2	10
Moriz, T.	—	—	—	—	1	6+3	—	—	8	5	23
Audele, T. de ? of Madingley	—	14	—	4	—	—	—	—	—	—	18
H. Blankpayne of Girton	—	—	2	—	[3]						2
quondam B. Payne Houdlo, Attegrene of Girton & Howes	—	[1+3]	—	—							[quondam 9]
R. Baldistone of Cotes, quondam owners of Cotes [] = quondam owner.	— —	— —	— —	— 2	[1] —	— 2	— —	— —	— —	— —	4 [quondam 5]
Others of Cambridge	—	—	—	—	—	—	3	—	2	—	5

157

THE WEST FIELDS OF CAMBRIDGE

AREA III

OWNER	Quondam Owner	MIDDLEFIELD					LITTLE				Total
		(end of) 3	4	5	6	7	(end of) 15	(end of) 16	7	
St. John's Hospital		1+4	2	—	—	1	1+2+2 (4+1)	1	6	20 [5]
Prior of Huntingdon		—	—	—	1+2	—	—	—	—	3
Prior of Barnwell		—	—	—	—	—	11+6 occ 4+1 } in 1 block	—	—	22
Nuns of Beche		—	—	—	1+2	1+5	1	—	—	10
University	(N. de Thornton)	—	3	—	—	—	—	9	—	12
C.C.C.	G. Seman T. de Cant. }	5	—	—	—	1+2	3+2	7+2+3	4+2	23
Moriz, S.	q. patris sui others }	} 2	} 2	—	} 1	} 1	} 1	—	} 3	4 } 6
Moriz, T.		—	—	—	—	—	3	—	—	3
Brigham Robt		—	9	—	—	—	—	—	—	9
Roger de Harl.	q. Wm Warde	—	—	13	—	—	—	—	—	13
Busshell, G.	S. de Refham	—	—	10+2	—	—	10+gore	—	—	22½
Thomas de Audele		—	2	—	—	—	2	—	—	4
J. de Cotton J. de Weston }	J. de Comberton	3	—	—	8	—	—	—	—	3+8
Others	Tuillet & J. de Berton }	—	—	—	1	2+2	—	—	—	1+2+2
J. Pilet q.p.s.	holding mainly in Newnham }	—	—	—	—	—	1+2	—	—	3
N. Crocheman		—	—	—	—	4	—	—	—	4
R. Martyn		—	—	—	—	—	3	—	—	3

APPENDIX E

AREA IV		LITTLE FIELD						
OWNER	Quondam owner	Q	1	2	3	4	5	6	Total
Hosp. S. J.	—		1	2+2+2 +3 w. gore	5+1 gore+ 6 w. 3 butts	1	10 butts	14	c.43
Prior de Hunt. [White Canons]		[6]				5			5 6
P. de Bernewelle	—		—	—	—	10+15+3	—	4+5	37
Cant. in S. Sep.	[T. Tuillet]		4	2	—	—	—	—	6
University	[N. de Thornton]		—	—	4 w. gore	—	—	—	4½
Corpus Christi College	T. de Cant.		—	3	—	—	1	—	6
	J. Poplyngton & others		—	—	—	—	1+1	—	
S. Morys	q.p.s.		1	3+2+2	—	—	—	—	11
	Others		2	1					
T. Morys		3+1+6+2	—	4	—	1	—	—	17
R. Tuillet	—		—	—	—	—	—	10	10
R. de Harleston	R. Tableter		4+6	2+1	—	—	—	—	13
J. Barker	N. Bradenasshe			1+1	4	1	—	—	7
N. Crocheman			—	—	3	—	—	—	3
J. Pilet	q.p.s.		3	—	4	—	1	—	8
J. de Weston	S. de Morden		—	—	—	—	—	1	1
T. Audele			—	—	10	—	—	—	10
R. Thacsted	Wm. Thacsted		—	—	—	11	—	8 butts	c.15

159

THE WEST FIELDS OF CAMBRIDGE
AREA V   CARMEFIELD

OWNER	Quondam owner	Q 1	2	5 (E. part) +6	7	13+ 14 (S. parts)	Total
St. John's Hosp.		5+1+2	2+1+1	2 headlands	1+1+1+1 +2+1	1+1+ (in a block) 20+6	22+ 20+6
Nuns of Beche		—	—	—	1+1+1+1 +1	—	5
Cant. B. V. M. in S. Pet.	[Sabina de Eylesham]	—	—	—	7+2 gores	—	8
University	[N. de Thornton]	—	—	2	4	—	6
C.C.C.	Thos. de Cant. G. Seman	—	2	5	1+3+1	2+1+1 balk	15½
	J. de Popylngton Others	—	—	1+2	1+9 butts	—	8½
S. Moriz, sen.	q.p.s.	—	1	3	2+3+6(4)	3	18
T. Moriz		2	—	—	1	2+1+1	7
N. Crocheman		—	1+1	6+5 + balk	—	—	13¼
J. Pilet	q.p.s.	—	2	2	—	1+4+6 + 3 butts	16¼
Ric Martyn	q.p.s. Joh. Martyn	2+1	1+2 (1)		—	—	19
	Others	—	2+2 1	—	4+1+1	—	
G. Wardeboys		—	6	—	—	—	6
Ric de Arderne	Matilda Hichin Others	5+2 —	— —	— 3	—	—	10
J. Barker	Nich. Bradenasshe	—	—	6	—	—	6
R Barber	R. Bolour	—	—	1+2	—	—	3
J. de Weston	S. de Morden Others	—	—	1	1+1	—	3
Henry de Beche		—	2	—	1+1	—	4
Adam de Kingston	q.p.s.	—	—	3	—	—	3
W. Aleyn de Bokesworth		—	—	—	1	—	1
R. de Harleston		—	—	—	—	1	1
Others		1	1	—	—	—	2
	No. of Selions Total	21	28	43½	63	27½+20+6	

INDEX

Acreages
   Maitland's calculations, 26
   Measured and estimated, 12, 15-6
   in Terrier, 9, 16
Ailgar the Noble, 19, 60, 73
Aldermanhill, 45, 47, 49, 72, 74, 118-9, 139-40, 145, 152
Aleyn, Wm., de Boxworth, 110, 132, 134, 138, 160
Alleine, Wm., 23
All Saints by the Castle, Church of, 7, 62, 75-6
All Saints in Jewry, Church of, 76
Amy, John, de Hitchin, 136
Aratral curve, 15, 53-4
Ardern(e), Richard de, 7, 75, 127-8, 131, 160
Ashton House, 66-7
Assarts, 26, 29, 38-9, 41, 54-5, 72, 74
Asshemannes Croft, 53
Assh(e)wyk(e)ston, 30, 36-7, 52, 60, 88, 149
At(t)(e)church(e) (Attecherche), Thomas, de Howes, 29, 50, 90-2, 94, 99-100, 111, 120, 139, 143
Attegrene, John, 91, 112, 157
Audele, Thomas de, de Grantchester, 74, 104-5, 108, 112, 116, 122, 125, 134, 157-9
Audre, Roger, 145
       Thomas, 146
Aunger family, 29

Backs (Backsides), 32, 43, 80-7
Baldeston, Roger de, de Cotes, 105, 112, 157
Baldwyn, John, 138
Balks, ix, 2-5, 17, 19-25, 32, 36-7, 54, and 88-138 *passim*
Barbour, Robert, 131, 160
Barker, John, 74, 124-5, 130, 131, 159-60
Barkersaker, 43, 46, 137, 155
Barking, Walter de, 146
Barnwell, 50, 62, 70, 144-6
   Field, 145 *see also* Fields
   Liber Memorandorum de, 60, 67
   Prior of, 17-8, 21, 24, 30, 40, 42, 54, 59, 62, 74-5, 92-3, 99-105, 107, 110-121, 125-6, 157-9
   Priory, 7, 47, 76, 78, 150, 152
   Sylvester de, 145
   Tithe Books of, 27, 29, 74, 76, 79, 90, 116, 119
Barton (Berton), John de, 89-90, 95, 101, 106-7, 109, 115-6, 119, 130, 156, 158
Barton Way, ix, 36, 38, 43, 71, 86, 107-8, 119, 122-7, 139, 141, 145, 147, 151, 153-4, *see also* Roman Road
Basketmaking, 147
Bateman, Richard, 145
Beche,
   Henry de, 7, 128-9, 132, 135, 160
   Nuns of, (Denney Abbey), 8-9, 30, 75, 89, 91, 93-4, 99, 101, 106-7, 109-111, 115-17, 120, 122-3, 129, 132-4, 136, 142, 156, 158, 160
   Sir Robert de, 144
Beck Brook, 34
Bekeswell (Bokeswell), Wm. de, 94, 101, 107, 111, 115, 118, 122-4, 134, 145
Benefactors, 75-6
Benelond, 47
Benet College, *see* Corpus Christi College
Benjamin, the Jew, 69, 144
Beresford, Prof. M., 15
Bernard, John, de Bernewelle, 145
Billingford, Richard (Master of CCC), 9
Bin Brook, 17, 32, 34-8, 45-6, 48-50, 80-4, 86, 123-7, 130, 133-6, 140-2, 145, 148, 151-5
Birmingham, Peter de, 146
Blackaker, (le), 47, 89, 119, 149, 152
Black Death, 76
Blackmore (Blackymore), 2, 14-15, 27, 37, 140, 142
Blackmore Field, 2, 15, 26, 29, 43, 54
Blancpayn,
   Henry, 91, 105, 157
   John, 92, 100, 104, 111
Blangernon, Baldwin, 61-3, 70, 75, 82, 151
   John, 105
   Thomas, 119
Bluntisham, Robert de, 121
Bokenham (Buckenham), 10, 99, 101-2, 108, 116, 122, 134, 136
Boll(e), Thomas, 7-8, 10, 32, 89-90, 94-6, 98, 105-6, 156
Bolour, Robert, 130-1, 160
Botwright, Dr. John (Master CCC, 1443-1474), 8
Brademere, 145
Bradenash, Nicholas, 124-5, 130-1, 159-60
Braderussh (Moor Barns), 16, 34, 46, 54, 100-2, 139, 150
Bradley, Simon de, 146
Bradshaw, Henry (University Librarian), 1, 7
Brady, Robert (Master of Gonville and Caius College), 66
Brandon, Roger de, 137
Branton, John, 131
Brembilfurlong, 19, 96-7, 150
Bridge Street, 9

THE WEST FIELDS OF CAMBRIDGE

Brigham (Brygham), Robert, 74, 90, 107-8, 134, 136, 156, 158
Brittany, Count Alan de, 70, 154
Broad, 14
Brook, (le), 45-6, 110, 145, 155 *see also* Bin Brook
Brunne, John (Receiver of rents, CCC, temp. Richard II), 9
Brunneforthdole, 19-20, 74, 109, 152
Buck, S., *Prospect of Cambridge,* 21
Buckenham, *see* Bokenham
Building, 31-2, 142
Burdeleis family, 34
Burial Mound, 44
Burton, William, 107
Burwash (Burghassh), Sir Bartholomew, 121, 153
Burwash, manor of, 73
Bushel, Geoffrey, 108, 116, 158
Butcher's Close (Mortimer's Close), 67, 155
Butcher's Croft (Mortimer's Meadow), 42
Butebroc, 46, 153
Butler, Jacob, of Barnwell Priory, 11
Butts, 2, 15, 31, 38

Cambridge, Corporation of, 23, 42, 56-7, 64-6, 85-6, 140,
  De, family, 74
  Field, 48-9
  Sir John de, 64, 70
    Cartulary of, 50, 69
  Thomas de, 17, 89-101, 103-5, 107-12, 114-7, 119-27, 129-34, 136, 156-60
  Watercourse called, 81
Canons, White, of Sempringham, 91, 94, 123
Carbonel, William, 146
Carme-
  dole, 19, 46-7, 67, 77, 138, 154-5
  dolefurlong, 77, 136-7, 155
  field, ix, 9-10, 22-3, 25, 27, 38-9, 41, 46-9, 67, 73-6, 82-4, 127-38, 139-42, 153-4, 160
Carmelites, 47-8, 67-8, 73, 145, 154-5
Castle,18, 38, 60, 149
  End, 30, 88, 149
  Hill, 7, 13, 35, 38, 42, 47, 49, 51, 56, 58, 60, 70, 73-5, 82, 149, 151
  Lane (Barton Way), 32
  Mound, 36, 52,
Causeway, 90
Caxton, Jeremy de (Sheriff), 144
Cayley (manor), 56
"Celtic Fields", 52
Centuriation, 53
Chalkpits, 143, 145
Chalkwell (dole) (croft), 106, 139, 145, 151
Chantries (all of BVM)
  Coton, 118
  St. Clements, 98, 102, 104, 110, 116, 118, 120
  St. Mary in the market, 90, 92, 100

St. Peter (Little St. Mary's), 76, 129, 132, 160
St. Sepulchre, 79, 90-2, 94, 97, 99, 102, 104, 116-7, 119, 121, 123-4, 131, 156, 159
Unspecified, 89
Chapman, 128
Charlock (Karlokaker), 47, 119, 152
Charner, John le, 146
Charters, 38, 45, 49-50, 70, 144-5
Chekker (le), 47, 104, 150
Chester, Ranulf, Earl of, 144
Chesterton, 1, 9, 38, 41, 44n, 56-60, 62
  William, Hugh and Sir Robert de, 59
Churchill College, 35
Civil War, 52
Clare College (Clare Hall), 8-9, 22, 31, 56, 90, 93-4, 96-7, 106-8, 111, 115, 117-8, 126, 129-36, 142
Clarkson Road, 30
Clay (le), 41, 49, 145, 151-2
Claypits, 29-30, 47, 88-9, 143, 149
Clint, 35, 44-5, 153
  -haveden, 17, 44-5, 50, 153
  -way, 17, 21, 35, 37, 44-5, 50, 122-3, 139, 141, 153
Close, 31
Coe Fen, 42
Cole, John de Barton, 10
College Field, 48, 154, *see also* Carme Field
Colville, Henry de, Sheriff, 61
Colys Cross, 37, 47, 123, 153
Comber, John le, 146
Comberton
  Geoffrey de, 146
  John de, 74, 95, 97, 100-1, 103, 107-9, 130, 133-34, 158
  Ralph de, 145
  Reginald de, 145
  Robert de, 129-32
  Thomas de, 17, 103, 105, 109, 114, 117-8, 138
Common Rights, 17, 23-4, 32, 57, 81-2, 85-7, 141
Compass Points, 13
Conduit (le Cundyt), 9, 47, 99, 101, 126, 150
  Head, 99, 139, 148, 150
  Wm. Atte-, 121, 135
Corpus Christi College (Benet College), 1, 7-8, 15, 23-4, 30, 32, 40, 56, 64, 67, 70-1, 74-5, 82, 85, 89-127, 129-137, 142, 150, 156-60
Costiseye, Reginald de, 146
  Roger de, 134, 136
Cotesweye, 43, 108, 115-6, 118, 139-40, 151-2
Cotesweyende, 49
Coton (Cotes), ix, 21, 29, 34-6, 39-40, 45, 53, 60, 74, 91, 112, 118, 147, 151-3
  Fields of, 112, 114-5, 118, 120, 122
  John de,¹ 17, 23, 102, 109, 114, 117-8, 136, 138, 158

INDEX

-path, 77, 85, 124, 129, 132, 151, 154
Cottenham, 19
  Andrew de, 129, 136
Cotton (Hall), 56
Crocheman, Nicholas, 74-5, 109, 117, 122, 128-9, 131-2, 158-60
Croft, 31, 38, 52-3, 96
Cropping 28, 47, 53, *see also* seasons
Crosses, 36-7
  Barton Cross, 139
  Colys Cross, 37, 47, 123, 153
  Drawings of, 9
  High Cross (in St. Neots Way), 21, 29, 35, 37, 40-1, 49, 104, 112-3, 139, 142-3, 152
  Hunnell's Cross, 37, 108
  Old Newnham Way cross, 37, 128, 140
  Stump Cross, 37, 139
  Stouping Cross, 37, 94
Cuckoo's, 66
Cultivation 28, 53, *see also* seasons
Cupitt's Crofts, 31, 142
Custes Balk (Custos Balk), 22, 36, 39, 46, 53, 77, 129-30, 132, 134, 139, 141, 154
Custes Weye, 125-6, 154
Cuttedrod, 48, 95, 150
Cutting, 94-95, 98

Dale (le), 19, 45, 60, 152-3, *see also* Deddale
Daleway (Weiebalk, Daleride, Deddale, Dedmannisdale), 21, 45, 118, 120, 152-3
Dame Nycholeshythe, 145
Danes, 55
David, Earl, King of Scotland, 65, 68-9, 72, 144
Ded(d)ale, 13-4, 21, 31, 35, 39, 45, 49, 120-1, 139, 141, 147, 152-3
Dedmannesdale, 45, 153
Dedmannesweye in Grantchester, 45, 153
Dedole Field, 26, 45, 48
Dene, William, Skinner, 107
Denney *see* Beche, Nuns of,
Dickins, Professor Bruce, 44, 44n, 46n
Ditton, Fen, 22
Dixy, Laurence, 145
Dole *see also* Porthorsdole, etc, 17-9, 70, 72-3
Domesday, 26, 34, 50, 72
Dovehouse Hill, 47, 77, 128, 154
Doy, Reginald, 144
Drainage, 41-2
Drakes Spring, 151
Duck End, 44
Ducksmere, 39, 96-7, 150
Dukdole, 19, 30, 42, 142, 148, 152
Dukpit, 41, 152
Dunning, 18, 51, 57-64, 69, 82, 149, 156
  Adam, 63, 68-9
  Eustace, 61-2

Hervey Fitz Eustace, 47, 56, 58-9, 61-3, 68-9, 144, 154
  Joan, 64, 145-6
  John, 64, 145
  Leonius, 69, 144
  Richard, 51, 69, 145-6
  Robert, 69, 100, 105, 145-6
Dyer, M., 114

Edes, John, 145
Edith the Fair, 70, 154
Edwin's Ditch (Edwin's Brook), 21, 31, 34-6, 42, 45-6, 49-50, 74, 119, 121-2, 124-5, 139-42, 147-8, 152-3
Eldenewenham, *see* Newnham
Elis, Thomas, 10
Elms, 147
Elyot, Master Adam, 145
Emparking, 84
Enclosure, 27, 28-32, 52, 84-7, 142
  Parliamentary, 31, 36, 57
  First Draft Map, 29
Endelesseway, ix, 20, 27, 36, 40-1, 43, 114-5, 139, 142, 152
Erbeer (le), (le Hearne), 48, 108, 139, 151
Erlesdole, 19, 47, 73, 121, 153
'Every year land', 31, 140
Exchange, 116
Eyre, John le, 146

Fadersoul, Alice, 121
Fen Ditton, 22
Fens, 55
Fields, 2, 26-8, 35, 37-8, 48-52, and 88-138 *passim*
  Acreage, 27
  East (Barnwell) Fields, 2, 10-1
  Extent of, 29
  West Fields, 2, 10-1, 18, 30, 32-4, 41
Finberg, Professor, H. P. R., 25
Finding code, of Bursar CCC and University, 8-9
Fishponds, 80, 84
Flax, 47
Fodder, 24
Foukes, Richard, 128
Franciscans, 47, 150
Fraternity of the Holy Sepulchre, 79
French's Mill, 149
Froshiscroft (Freshcroft, Hunnell's Croft or Close), 20, 31, 42, 47, 106, 151
Froshlake Way, 36, 47, 66, 83, 129, 131-2, 137, 140, 154
Fulbrook, 35
Furlong, 2, 12, 25-6, 39, 86, and 88-138 *passim*
Fynne, J., 10

Gagges Aker, 134, 155
Garrett Hostel, 32, 80
Garrison, 54

163

Gildenaker, 47, 97, 115-6, 150, 152
Girton, 21, 29, 34, 41, 44, 52, 91, 105, 149-50
    Field, 100, 139, 150
Godivesdole (Goidesmedole), 19, 46, 60, 73, 122-3, 153
Godwinesrode, 46, 152
Godyn, John, chaplain, 133, 155
Godynys Buttes, 133, 155
Goggying, Bartholemew, 69
    John, 146
Golderyng, John, 146
Goldsmith, Robert, 100, 102, 110-1, 117
    Walter, 121
    William, 153
Gonville and Caius College, 10, 49, 64, 66-7, 106, 108, 116, 119, 122, 127, 134
Gonville Hall, 66
Goredaker, 48, 98, 102, 117, 132, 150, 155
Gores, 2, 15, 38
Grange, The, *see* St. John's College Farm
Grange Road, 22, 36, 154
Grantacaestir, 56
Grantasaete, 35
Grantchester, 34-6, 47, 50, 57, 70, 75, 128, 140, 142, 153-4
    Mill, 70, 154
    Road, 36
    Street, 140
Gravel, 44
Gravel pits, 8, 29-30, 36, 47, 86, 93, 98, 100, 106, 143, 148
Gray, A., 28
Greenplat (Grenesheld), 30, 44, 49, 93, 149
Griffyn, 111, 116, 126
Grithow, 9, 30, 44, 60, 99, 149-50
    Field, ix, 3, 14, 22-3, 25, 27, 31, 38-9, 41, 44, 47, 48-9, 52, 60, 73, 88-105, 139, 143, 149, 152, 156-7
    Hill, 30, 93, 99, 139, 151
    Path, 30, 36, 44, 93, 99, 149
    Way, 36, 44, 73, 89, 93, 95, 98, 139, 149-50
Guy of Barnard Castle, 61
Gyboun, John, 7

Hammond's Map of Cambridge (1592), 31, 66
Hardy, Andrew, 128
Harleston manor, 56
    Roger de, ix, 20, 32, 64, 74, 90-103, 106-8, 110-20, 122-4, 127, 129, 130, 133, 135-7, 142, 156, 158-60
Haveden, 17
Hawthorn, 148
Hay, 54-5
Headland *(forera)*, 2, 16-7, 23, 55, and 88-138 *passim*
Hearne (le Erbeer), 48, 108, 139, 151
Hechyn, Matilda, 128, 160
Hedges, 21, 31, 148
Hervey fitz Eustace, *see* Dunning
Hervey, Michael son of, 144

Herwardaker, 46, 153
Hides, 3
Highgable (Haggable), 66, 144
Hinton Moor, 145
    Way, 35, 145
Hitchin, Andrew de, 146
Hodilow Farm Terrier, 15
Hor Hill (Hores Hill), 47, 89, 143, 149
Hospital of St. John, *see* St. John's Hospital
Houdlo (Howdelowe)
    Roger, 100, 104, 157
    Simon, 116, 120
Howes, 21, 29, 33-4, 36, 39, 41, 44, 52-3, 74, 91, 146, 148-9 *see also* Attechurche, Thomas, de Howes.
Howescroftesende, 44, 49, 149
Howeshedge, 149
Huberd, Robert, 69
Hugh, John son of, 146
Hundred Rolls (Rotuli Hundredorum), 47-8, 57n, 67-8, 76, 152
Hundreds, 38, 55, 60
Hunnell's Close, 42, 93, 106, 139, 151, *see also* Froshiscroft
Hunnell's Cross, 37, 108, 151
Huntingdon, Earldom of, 69
Huntingdon, Honour of, 144
Huntingdon, Prior of, 18, 74-5, 82, 88, 91-5, 97, 99-100, 102, 105, 106-7, 109, 111-2, 116, 125, 129-30, 134-6, 156-9
Huntingdon Road (Way), 13, 21, 23, 28, 34-5, 37, 44, 84, 88, 90, 97, 149
Hynde-Cotton family, 29

Indebtedness, 63
Iuxta, 3

Jekke, Thomas, 52, 91-2, 100
Jesus College 8, *see also* St. Radegund, Nuns of
Joachim, 146
John, King, 68-9
Jordana, 146

Karlok, Richard, 152
Kerridge, Dr. E., 4, 15
Kiln, Stephen at ye, 146
King's College, 82, 84-5, 121, 135, 140
    Green, 155
    Way, 155
Kynchmade, 34, 46, 99-100, 147, 150
Kynxton (Kingston), Adam de 131, 160

Lampeaker (le), 47, 128, 154
Landownership, 72-6, 156-60
Lane
    Gilbert in the, 120
    Robert in the, 119
Laurence, Richard, 146
Lavenham, William de, 90-2, 94, 99-100

INDEX

Leckhampton House, 36
Leys, 139
Lincoln, Earl of, 73, 153
Lingwood, William, 96
Linrode, 47
Litilmer (Litilmor), 43, 49, 90, 149
Little Carme Field, 26, 45, 48, 50
Little Field ix, 13, 18, 22-3, 25, 37, 41, 48-50, 52, 73-4, 82, 123-7, 141, 153, 158-9
Loggan, D., *Cantabrigia Illustrata,* 23, 36, 147
Lolleworth, William de, Alutarius, 7, 90, 93, 96-7, 101, 106, 109, 111-5
London, Richard de, 97
London Waye, 145
Long, Robert, 14, 54, 92-3, 95-7, 99, 101-2, 104, 106, 110-1, 117-8
Long Balk, 22, 36, 39, 53, 77, 83, 129, 131-2, 134, 136-7, 140, 142, 151, 154
Long Green (Meadow), ix, 24, 32, 37, 42-3, 56, 82-7, 135, 140, 155
Lordship, 57-72
Lovell, of Chesterton, 89, 116, 130, 135-6
Low, The, 120, 153 *see also* Deddale
Lucas, R., 10
Lykylt Grene, 43, 119, 153

Madingley, 17, 29, 34, 45, 151
  Hill, 41
  Moor, 41, 44, 49, 102-3, 105, 143, 150-1
  Road, *see* St. Neot's Way
Maitland, F. W., v, 1, 3, 13-5, 18, 25, 26, 29, 41, 45, 48, 50, 54, 57-9, 63, 63n, 66, 68, 70, 73, 76, 79, 82, 86, 152
  see also Township and Borough
Malherbe, Michael, 67
Malting(house) Lane, 47, 154
Manors, 4, 57-8, 66-8
Manure, 54-5
Maps, 5, 29, 54, 66, 80-1, 85-6
Marbilthorp, John de, 145, 146
  Thomas, 90-1, 95-6, 107, 111, 119
Mare Way, 35
Mariota de Grantesete, Everard fil, 75
Marsh, 52
Marshall, John, 108
Martyn, Avicia, 146
  Hervey, son of, 144
  John, 116, 128, 134, 136, 146, 160
  Richard, 7, 74-5, 116, 127-8, 130, 132, 137, 158, 160
  Robert, 128-9
Masters, Robert, 6, 11
Matfrey, Henry, 145
Merton College, Oxford, 8, 14, 18-9, 29-31, 42, 44, 48-9, 56-7, 64, 73, 84-5, 91, 93, 98-100, 102-4, 119, 121-2, 133, 135, 142-3, 156
Merton Hall (the Stone House), 37, 57, 61-2, 82-3, 88, 145
Middle Field (Medylfelde etc.), ix, 14-5, 18-9, 22-3, 25, 27, 30, 38-9, 44-5, 48-9, 53, 72-4, 78, 105-23, 141, 150-1, 153, 156-8
Midwife, Malina, 146
Miller, Sayer, 146
Mills, 46
  Grantchester, 70
  King's, 65, 80
  Newnham, 33, 36, 46, 65-6, 70, 75, 77, 80, 83
Milnedampondes, 77, 154
Milneway (Milnepath), 15, 36, 99, 149
Milneyerde, 154
Moor, 29, 38, 43, 72
Moorbarns (Braderussh), 21, 29-30, 35, 41, 45, 105, 115, 139, 151
Moorden, Simon de, 126, 130-2, 135, 159-60
Mordole, 19, 21, 40, 100, 150
Morhyl, 49, 150
Morley, Sir Matthew de, 144
Morris (Morice, Morys), 8, 64, 74, 150
  Matilda, 146
  Nicholas, 122
  Richard, 97, 130, 133
  Stephen junior, 7, 42, 93, 98, 100-1, 106-7, 111, 114, 117-8
  Stephen senior, 8, 10, 89-90, 92-5, 97-101, 107-9, 112, 114-20, 122-4, 126, 129-34, 136-7, 156-60
  Thomas, 40, 44, 89-93, 95-101, 103, 107, 112-4, 116-20, 123-5, 130-2, 134-7, 142, 156-60
Mortimer, ix, 18, 50-1, 63-72, 88, 100, 106, 119, 126-7, 151-2, 156
  Close, 67, 135
  Dam, 154
  Documents from Gonville and Caius College, 144-6
  Manor, 10, 50, 65-70, 154-5
  Meadow (Butcher's Croft), 42, 67, 77, 82-3, 128, 137-8, 154-5
  Robert, 67, 155
  Sir Constantine de, 67, 69, 145
  Sir Robert de, 65, 68-70, 72, 144
  Sir William de, 69, 144
  William, 67
Mounford, Brice, 146
Mount Pleasant, 149
Mulleweie, 149
Mundes lond, 123, 153
Muscroft, 31, 37, 47-8, 52, 71, 151
Muscroftfurlong, 38, 105, 151

Nakedole, 19, 97, 150
Newnham (Eldenewenham), ix, 22, 33-4, 39, 41, 43, 48-51, 53, 63, 66-8, 70, 72, 74-5, 78, 82, 128-9, 144-6, 150, 154-5, 158
  College, 47, 154
  Crofts, ix, 31, 33-4, 50, 52-3, 75, 77-8, 127, 131, 154
  House, 66-7
  Manor, 66

THE WEST FIELDS OF CAMBRIDGE

Mill, 65-7, 70, 80, 82, 145, 154
Old Newnham Way, 31, 35, 37, 43, 47-8, 53, 75, 128, 130, 140, 154
Nicholas, Vicar of St. Clement's, 107
Nightingale's, 66
Niket, Richard, 100
Nobys, Peter, Master CCC, 10
Normans, 56, 60
Norton, John, 103-4
Nutters Yard, 84

Oaks, 148
Observatory, 30
Open Fields, See Orwin
Orwin, C. S. & C. S., 1, 4, 19, 22, 24
Oxenford, Joan de, 146

Panton, 8
Parishes, 33-4
  boundaries, 53, 78, 81
Parker Library, 2
Parker, Matthew, Master CCC, 10-1
Parker's Piece, 86
Parson, Andrew, B. D., 11
Payne, B., 157
Peasants' Revolt, 1
Peperdole, 19, 47, 99, 150
Pestour, Henry, 146
Pestour, (le), 50, 75
Peterhouse (Domus Sancti Petri), 76-7, 127-9, 131
Picot, 81
Pilat (Pylat), Hervey, 146
  Michael, 146
  William, 144
Pilet, John, 74, 116, 122-3, 125, 129, 131, 135, 137, 158-60
Pittok, Hugo, 93-4, 106, 117, 120, 122-3
  John, 90, 101, 106, 114, 117-8
  William, 145
Place-names, 43-51, 149-55
Ploughland, 54-5
Poplynton, John de, 75, 126, 130-2, 135, 159-60
Portebrigg (Portbridge), 35, 37, 42, 50, 123-4, 126, 128-30, 139-41, 147, 154
Portefield, 9, 48-9, 56, 74, 79, 145, 151
Porthors, John, 145, 152
Porthorsdole, 19, 46, 109, 142, 152
Postan, Professor M. M., 3, 24, 55
Pound Green, 37, 52
Priorsdole, 73, 94-5, 98, 149-50
Psychewellewaye, 145

Quarentena, see Furlong,
Queens' College, 8, 80, 87
Queens' Road, 42

Ralph, Geoffrey son of, 144
Ramsey Abbey, 79

Rawlins, Margaret, 38
Rawlyn(s), 99, 101, 116, 122, 134
Reaney, P. H. 44
Rede, 128, 130-1
Redhod, John, 89, 95, 100, 107-9, 112, 115, 117-8, 121, 156
Reepham (Refham), John de, 100, 108, 116
  Simon de, 146, 158
Ridge, *see* selion
Roads, 13, and *passim*
Robinson, Wilton, of the Birdbolt in Barnwell, farmer to CCC, 11
Roman remains, 41, 52
  road, ix, 34-5
Roper, William le, 146
Round Church, *see* St. Sepulchre
Royal Commission on Historical Monuments, 33
Russell, Roger, 132, 138
Rye, Robert, 52, 91

Sabina née Brithnoth, vid. John de Aylsham, 76, 160
Sabyn, Robert, of Grantchester, 145-6
St. Benet's Church, 9
St. Bernard's Hostel, 8
St. Botolph's, 8, 67, 78-9, 89, 122 and 88-138 *passim*
St. Catharine's College, 34
  Hall (Katerine Hall), 94, 118
  Willows, 147
St. Clement's 76, 145
St. Edmund's Chapel, 145
St. Giles (Egid), 7, 47, 57, 76, 90, 114, 121, 128, 151, and 88-138 *passim*, 157
  Acre, 79, 90, 106, 151
  Rod (Gillisrod), 79, 107
St. John Baptist's parish church in Milnestrete, 126, 135
St. John's College, 8, 32, 56-7, 64, 84-6, 108, 122, 140, 151, 154 *see also* St. John's Hospital
  Barns, 37, 52, 71, 83, 105, 127, 148, 151
  Claypits or Chalkpits, 143
  Close, 32, 139
  Ditch, 80-1
  Farm (Grange Farm), 36, 71, 151
  Meadow, 84-6
  Terrier, 23, 53, 85
  Wilderness, 42, 82
St. John's Hospital, 64, 66, 68, 84, 142, 146
  Lands, ix, 17-8, 45, 47, 52, 59, 68, 70-2, 75, 78, 83-4, 88-100, 102, 104-12, 114-7, 119-38, 142, 150-1, 154-60
  Sheepcote (le Bercarye), 71, 107-8, 126, 142, 151
  Terrier rolls of, 116
St. Mary and Corpus Christi, Guild of, 150
St. Mary the Less (without Trumpington Gates, St. Peter without Trumpington Gates), 67, 76-8

166

INDEX

St. Mary in the Market, 92, 100
St. Michael, 67
St. Neot's Way (Madingley Road), 13, 20, 22, 28, 34-8, 41-5, 49, 60, 73, 84, 93-4, 101, 103-7, 110, 112-3, 127, 139, 141-3, 149-52
St. Peter without Trumpington Gates (St. Mary without), 67, 76-8, 129, 132
St. Peter ultra pontem, 76, 81
St. Radegund, Nuns of, 59, 75-6, 78, 93, 96-7, 101-2, 114-5, 117, 119, 121, 123, 127, and 88-138 *passim* as tithe owners, 145, 157
St. Sepulchre's Church (Round Church), 76, 78-9, 89, 95, 108, 110, 115-7, 119-22, 125-6, 131-2, 134-6
Saladyn, John, of Barnwell, 145
Sale Piece, 25, 31, 38, 60-1, 88, 149, 156
Savery, Sir John, de Cotes, 112
Scalers, John de, 144
Scandinavians, 44-5, 54
Scroope, Lady Anne, 65
Seasons (cropping), ix, 22, 26-8, 31, 53, 139-43
Seebohm *(The English Village Community)*, v, 1, 3, 5, 14, 19
Selion, 2-3, 5, 8, 12-7, 23-4, 30-2, 38, 44, 72, and 88-138 *passim*
Seman family, 64, 74, 150
Seman, Geoffrey, 64, 89-91, 93-9, 101-4, 106-7, 109-118, 120, 126, 133-7, 156-8, 160
  Laurence, 145-6
  Robert, 91, 118, 144
Seman's Ditch, 43, 46, 99, 101, 139, 150
Sengham, William de, 75
Sengilaker, 90, 149
Sexton, John de, 144
Sharp, John, 128
Sheep's Green, 35, 42
Shelford, John de, 145
Shepecote, 108, 151
  Way, 37, 43, 74-5, 151
Shermannisrod, 46, 129, 155
Sherriffesdole, 19, 40, 47, 112, 152
Sherwynd, John, 109
  William, 121
Shift, rotational, *see* seasons
Shire-Town, 55
Slegge, 10
Smalemade, 36, 43, 49, 108, 151
Small Bridges, 82, 141
Snoryng, William, 125, 135
Social change, 62-4
Spaldyng's Close, 29, 47, 100
Sparewe, Walter, of Coton 145
Sparrows Croft, 118, 120, 152
Spicer, Julian le, 146
Spiteldole, 50, 145, 155
Sponyaker (le), 47, 89, 149
Stacked Acre, 98
Stephen, J., 10
Steresman, Robert, 145

Stockton, Simon de, 146
Stokes, H. P., 33
Stonigrund, 47
Stonydale, 145
Stothydes Acre, 95
Stoupendcruche(way), 36-8, 89, 94, 139, 143, 145, 149
  (furlong), 94, 98
Stow, Aubrey de, 138
Strips, 2-3, 14, 18
Stump Cross, 37
Sturbridge (Steresbregg) Chapel, 90-1, 96, 156
Sturmyn, Thomas, 135
Suthwynd, William, 121
Swynecroft 145

Tableter, Richard, 64, 91-3, 96-101, 106-7, 118, 120, 122-4, 130-1, 133, 135-7, 159
  John, 114, 117-8
Tangmere (Tanglemere), Henry de, 74-5, 129, 135, 138
Taverner, Geoffrey, 92
Templeman, John de Cotes, 118
Terriers, 22-3, 26
  Corpus or Cambridge, ix, 1-11, 15-9, 22, 24-7, 29-31, 42, 44-8, 52-4, 76, 78, 149
  Clare Hall, 22, 25, 30
  Jesus College, 48, 139
  St. John's, 16, 23, 27, 29-30, 42, 53, 139-43
Thaxsted (Thackstede), Richard, 126, 159
  William de, 118, 125, 159
Thomas fitz Joachim, 144
Thompson's Lane, 56
Thornton, Nigel de, chaplain, 76, 79, 158-60
Thorpe's Croft, 15, 46, 116-8, 152
Tithe 2, 7, 24, 76, 78-9, 88, and 88-138 *passim*, *see also* Barnwell, Tithe Books of,
  St. Botolph's, 105, 123, 127, 138
  St. John's, 108
Toft, John de, 89-90, 96, 105-6, 129, 156
  Henry de, 121, 146
Tolbooth, 65, 69, 144
Topography, notes in terriers, 139-43
Touche, John, 128
Township and Borough, v, 1, 3, *see also* Maitland, F. W.
Trees, 147-8
Trinity College, 86
  Close, 32
Troubelville, de, 18, 75, 156
Trumpington, John de, 146
  Milo de, 146
Trumpington Road, 70, 149
Tuliet, Richard, 32, 70, 93, 96-8, 105, 109, 115
  Robert, 93, 95, 115, 117, 120, 126, 159,
  Robert, son of Henry, 145

THE WEST FIELDS OF CAMBRIDGE

Thomas, 69, 144-5, 159
William, 79, 107, 109, 121, 145, 158
Tunmanisaker, 47, 94-5, 98, 150
Turnpike road, 31, 35, 37, 86-7, 151
Tylers, 134

University, 76, 86
  Archives, 8, 51, 76
  Farm, 15, 41, 47
  Lands, 51, 74, 76, 79, 108, 119, 122, 125, 131-6, 158-60
  Upper Town, 52, 56, 151

Vicar's Brook, 35
Victoria County Histories, 33
Villeinage, 3-4
Virgate, 3, 18

Walnuts, 148
Walter, Vicar of St. Clement's, 145
Wapentakes, see Hundreds
Warde, William, 74, 90, 92, 95-7, 100, 102-3, 108, 110, 115, 118, 120, 127, 129, 133, 156, 158
Wardeboys, Geoffrey, 53, 128, 160

Washpit Brook, 46, 150
Watercourses, 45-6, 80-5
Weiebalk (Daleway), 73, 118, 121, 152-3
Welde, Matilda de, 145
Welle, Alan de, 145
Weston, John de, 9, 93-4, 106, 108, 126. 129-36, 158-60
Wetherysfeld, 146
Whelpes Acre, 137, 155
White Canons of Sempringham, 8, 91, 94, 159
Whitwell, 34, 41
Willingham, 28
Willows, 147
Willows Brook, 16
  Close, 151
  Ditch, 20, 30, 34, 42, 46, 48, 106, 108-9, 116, 142, 147, 151-2
  Mere, 150
Wood, Mr., 8, 10, 90, 93-4
Wrangling Corner, 45
Wyntryngham, Thomas, 128
Wymak, John, de Cotes, 118
  Nicholas, de Cotes, 118

Youn(g), Richard, 98, 100, 118
  Roger, 11
Yslep(er), Roger, 93